Season at the Point

ALSO BY JACK CONNOR
The Complete Birder: A Guide to Better Birding

SEASON *at*

Jack Connor

Illustrations by Don Almquist

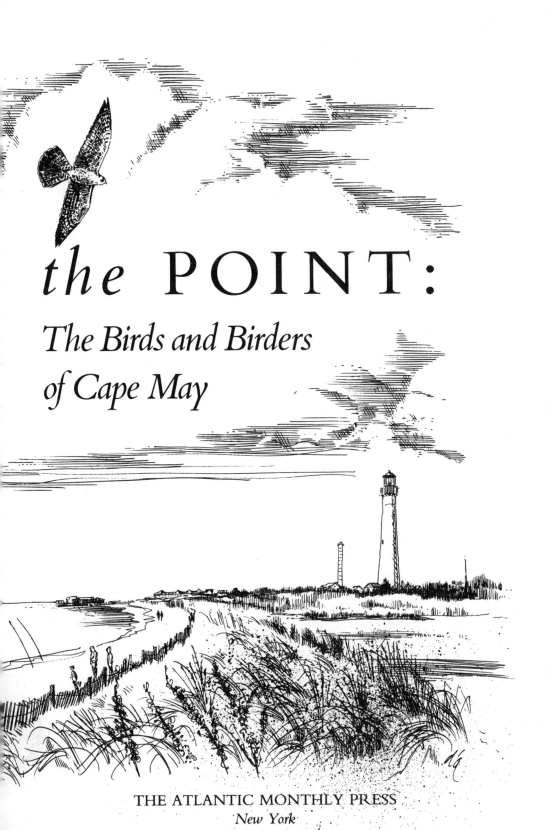

the POINT:

The Birds and Birders
of Cape May

THE ATLANTIC MONTHLY PRESS
New York

Lines from "Two Inch Wave" from *Relations: New and Selected Poems* by Philip Booth.
Copyright © 1986 by Philip Booth. Used by permission of Viking Penguin, a division
of Penguin Books USA Inc.

Grateful acknowledgment is made to David Sibley for the hawk illustrations appearing
in the Appendix, "A Short Guide to Cape May's Hawks." Illustrations copyright ©
1988 by David Sibley; reprinted by permission of Houghton Mifflin Company, Boston,
from *Hawks in Flight* by Pete Dunne, David Sibley, and Clay Sutton.

Published simultaneously in Canada
Printed in the United States of America

Library of Congress Cataloging-in-Publication Data
Connor, Jack.
Season at the point: the birds and birders of Cape May/Jack
Connor; illustrations by Don Almquist.—1st ed.

1. Hawks—New Jersey—Cape May Point State Park.   2. Birds—New
Jersey—Cape May Point State Park.   3. Bird watching—New Jersey—
Cape May Point State Park.   4. Cape May Point State Park (N.J.)
I. Title.
QL696.F32C66 1991   598'.916—dc20   91-9048
ISBN: 0-87113-475-6 (pbk.)

Design by Laura Hough

The Atlantic Monthly Press
19 Union Square West
New York, NY 10003

FIRST PRINTING

for Jesse

what is here, now,
is *here, now,*
beyond every knowledge
except our caring

—Philip Booth

Contents

CONTENTS

Illustrations

❖

ILLUSTRATIONS

❖

Chapter One

❖

DAY ONE

"HEY, THE STARLINGS HAVE SOMETHING," SAYS JEFF BOUTON, LIFTING HIS binoculars to his eyes. "See them balling up? *Harrier!* Going left, just right of the round-topped pine."

Two dozen starlings swoop down in a tight flock, not quite making contact with the hawk below. The harrier tilts, flaps once, flaps again, then wobbles left, low over the tree line. The starlings regroup and swoop once more. The hawk keeps on coming—wobble, flap-flap, wobble. The undersides of the wings flash cinnamon.

"Immature female," says Bouton.

As she reaches a pair of tall cedar trees, the starlings let her go. She crosses in front of the water tower, flaps past the lighthouse, veers left past the silo behind the lighthouse, and finally disappears over the red-shingled roof of St. Mary's-by-the-Sea, a retreat house for nuns that is the southernmost building in New Jersey. The hawk is heading west-southwest, across the fourteen miles of salt water at the mouth of the Delaware Bay.

Northern harrier over phragmites

"Now we're cooking with Crisco," says Bouton, lowering his binoculars. "That's raptor number one."

He is standing on a raised deck that is the hawkwatch platform at Cape May Point State Park. The houses and hotels of Cape May City, two miles to the east, shimmer in hazy silhouette under the rising sun. Dragonflies hover above the pond in front of the platform. Above them, martins and swallows loop through the air, squeaking and chirping. Laughing gulls *ho-hah* from the beach to Bouton's right; a wren sings from the woods to his left. Cars pull into the parking lot behind him steadily, and doors and trunk lids bang as beachgoers unload. "Have you got the sunscreen?" *"Jennifer,* wait for your mother!"

"Kestrel," Bouton announces. "Over the round-topped pine,

going left. Probable female . . . Definite female. Hold it—there's another. Two A.K.s coming, both females."

The two falcons follow the same route as the harrier, but they are smaller birds, pumping harder, flying more swiftly. The starlings do not pursue.

Bouton is barefoot, tan, and twenty-one years old. He wears a tight-chested T-shirt, baggy black shorts, a wristband of braided rope, and the tab from a pop-top soda can on a silver chain around his neck. The peak of his baseball cap is folded upward, like a bicycle racer's. His sunglasses hang from a red cord. "A punk birder," he calls himself.

His equipment is modest: a spiral notebook, a thirty-cent pen, a palm-sized VHS weather radio, a pair of 8.5×44 Swift binoculars, and a battered Bushnell spotting scope strapped loosely to an old gun stock by strips of fraying duct tape.

It is August 17 and the first hour of the first day of the thirteenth year of the Cape May Bird Observatory's Autumn Hawk Watch. Bouton will stand here scanning the skies from dawn to dusk seven days a week for the next three and a half months. Depending on weather, winds, the nesting success of the raptors that have bred to the north, and his own attention to the task, he expects to count between fifty thousand and eighty thousand hawks passing his position—more hawks in one season than most people, even most bird-watchers, will see in a lifetime.

The Point's census has averaged 66,000 hawks a season for the last twelve fall seasons, more than eighty hawks for each hour of observation time. No place in the Eastern United States can match these numbers. Hawk Mountain, Pennsylvania, the lookout with the next highest count, averages a third as many. For variety, no lookout in all of North America and few places in the Western Hemisphere compare with Cape May. Twenty-four species of hawks have been recorded at the Point, and seventeen of them are annual migrants, including two eagle species, three falcons, five buteos (red-tailed, broad-winged, red-shouldered, rough-legged, and Swainson's), and three accipiters. The two most cosmopolitan hawks, the osprey and the peregrine falcon, both of which occur on all continents but Antarctica, are seen in greater numbers in Cape May than anywhere else in the world. And hawks are only one part of the migration pageant on the Cape May peninsula. Eight owl species, seventeen sparrows, thirty-six warblers, and forty species of shorebirds are annual visitors. Western kingbirds and clay-colored sparrows from the West, black rails and purple gallinules from the South, lapland longspurs and white-winged crossbills from the North, Eurasian wigeons and Eurasian green-winged teals from Iceland and farther east—all are virtually expected birds in Cape May, though they are rarely seen elsewhere in New Jersey or in any other mid-Atlantic state. The list of "accidentals" recorded at the Point includes black-browed albatross, South Polar skua, Eskimo curlew, magnificent frigatebird, sooty tern, white-winged tern, chestnut-collared longspur, and dozens of other exotic species. When Bouton and two companions spotted a long-billed curlew flying over the hawkwatch platform on October 9 last year, the list of all bird species recorded in Cape May County reached 410, a total higher than the lists for most states in the U.S. and equal to the list for all of mainland Alaska, an area more than a thousand times as large.

"The curlew came right out of the gap," says Bouton, pointing, "calling all the way. Best nonraptor I saw all fall." "The gap" is a low row of phragmites and buttonbush beyond the pond in front of the platform that leads to a line of taller oaks, poplars, and tupelos bordering the South Cape May Meadows half a mile east. Right, or south, of the gap are two red-and-white radio towers and "the merlin sticks"—a cluster of dead trees where merlins often perch, sometimes feeding on songbirds and shorebirds they have caught. Farther right and south stands a powder-blue water tower and then the jumble of rooftops and steeples of downtown Cape May. The line of buildings ends at the Second Avenue jetty, which is hidden behind the grassy dunes that run around the inlet and up the beach back to the platform. The most prominent landmark on Bouton's right is a concrete bunker two hundred yards from the platform. It was three hundred yards from the water when it was built in 1942, equipped with four 155-millimeter artillery guns, and covered with sod. The Point's beach has been eroding for decades, however, and the sod and land around the bunker are long gone. Today a dozen fishermen are standing on the bunker's roof, their lines leading to the surf breaking against the bunker's wooden pilings, which were once forty feet underground.

"Falcons tend to come up the flight paths between the gap and the bunker," says Bouton. "The ones we get to see, at least. I'm sure we miss a lot of merlins and peregrines that fly down along the ocean paralleling Second Avenue and never turn this way. When they get to the jetty, they jump off right there, heading straight south. 'Delaware, here we come.' I've seen a few peregrines do it—on really clear days when I just happened to be scanning that way with the scope. But most days they're too fast and that's too far away."

Most other hawks—eagles, buteos, and accipiters—are first seen cresting the horizon north, or left, of the gap. Bouton's landmarks in this direction are a red-and-white radio tower; three prominent trees he calls "the round-topped pine," "the round-topped oak," and "the flat-topped oak"; a line of cedars; and finally the most useful directional guide on the platform's scene—a hundred-foot wooden tower eighty yards from the platform. It was a radio tower during World War II, part of the Coast Guard station at the Point, and each of its three guy wires still holds five insulator caps spaced at regular intervals. "They must have known back then there'd be a hawk-watch here someday," says Bouton, "so they never took it down.

That tower is about the best thing that ever happened to hawkwatchers in New Jersey. You tell folks there's an eagle out beyond the
bottom two insulators on the left-hand guy wire, and everybody's
on it in a second."

Why Cape May sees so many birds is largely a mystery, although two
obvious factors are its peninsular geography and its rich mix of
natural habitats. Hawks tend to follow "leading lines" on their migration paths—mountain ridges, river valleys, and lake, bay, or ocean
shores that both direct and restrict their flights. Cape May stands at
the meeting point of two such leading lines, the Delaware Bayshore
on the west and the Atlantic Coast on the east. Southbound hawks
following either coast are funneled to the Point, where they find
themselves facing the fourteen-mile overwater flight to Delaware.
Depending on their species and the winds of the day, they may depart
directly, backtrack up the Bayshore, swirl up high over the Point to
gather altitude for a long glide out, or put down for the night in the
woods at Higbee's Beach, Pond Creek, or elsewhere around the
peninsula. Hawks can still find food and protection in these woods.
Cape May County's human population has nearly doubled in the last
twenty years, and local conservationists despair about the future, but
at the moment much of the county remains relatively undeveloped.
Swamps, creeks, forests, orchards, pastures, and saltwater meadows
and marshes form a patchwork refuge for migrating birds of all kinds.
    But explaining the Point's hawk numbers by noting that the
South Jersey peninsula is a funnel and refuge does not do justice to
Cape May's special attractiveness to birds of prey, nor does it touch
on the puzzles behind the magnitude of the phenomenon. It is like
explaining that Henry Aaron holds the major-league record for home
runs because he was a well-conditioned athlete with quick reflexes,
or that Napoleon conquered Europe because he was an experienced
military man with a lot of ambition. "Funnel" and "refuge" are part

of every explanation of the Cape May hawk migration because they are the most identifiable components of the phenomenon. We know there must be other causes for the magnitude of the Cape May flight; we just don't know what they are. Really, Cape May's hawk migration has never been fully explained. At our present level of knowledge, it *cannot* be fully explained.

Run your eye down a map of eastern North America and you will note that a couple of dozen peninsulas along the Atlantic coast are oriented more or less in a southerly direction. All of them are migration funnels; many are green and undeveloped; all of them are good places to look for birds on an autumn day. Not a single one of them sees the number of hawks Cape May sees. Some can be dismissed as unworthy rivals, of course. Baccaro Point in Nova Scotia is obviously too far north; the majority of hawks nest south (or west) of the Maritime Provinces. Cape Sable at the tip of the Everglades is too far south; most hawks winter north of the Florida peninsula. Connecticut's Hammonasset Point and South Carolina's Cape Romain are both located on peninsulas which are apparently not large enough for a hawk flight to build—their funnels are too short. The peninsula that leads to North Carolina's Oregon Inlet is too thin, more straw than funnel. (Cape Hatteras, Cape Lookout, and Cape Fear, all farther south on the Carolina coast, are on islands, not peninsulas. Since the majority of hawks do not cross water unless forced to do so, island-based hawkwatches generally record only ospreys and falcons, the most seaworthy raptors.)

But other places make better, more intriguing comparisons. Why, for example, is Cape May's flight so much larger than the hawk flight at Point Lookout in southern Maryland? Point Lookout stands at the junction of the Potomac River and the Chesapeake Bay, both of which seem to qualify as potential leading lines. The St. Marys County peninsula is green and undeveloped, very similar in size to the Cape May peninsula, and at a latitude only one degree different from Cape May's. No one bothers to count hawks at Point Lookout, however. In fact, the hawk flight is so meager that (according to the records of the Hawk Migration Association of North America) no one has ever bothered to conduct a season-long count there. The Potomac is slightly smaller than the Delaware, and the Chesapeake is not the Atlantic. Does that make all the difference? Or could the shapes of the two peninsulas be a crucial factor? Most hawks coming south from Canada and New England are heading toward the Carolinas, Georgia, and the Gulf Coast—on a track that runs roughly southwestward, paralleling the Appalachian Mountains. The St. Marys County peninsula bends eastward against this track; the Cape May peninsula bends with it. Is that southwestward bend one of the reasons so many hawks are led to Cape May Point? Are they tricked by the promise of this right-handed turn—fooled into coming down onto the peninsula because it seems that it will carry them where they want to go? Would the Cape May flight be less spectacular if the peninsula were a mirror image of itself and Cape May Point, like Point Lookout, stood on the easternmost extension of the land? We don't know the answers to these questions. Do hawks have an internal compass that helps them orient during migration? Can they distinguish a southwestern bend from a southeastern one? We don't know. Are the hawks that reach Cape May there by choice, or are they lost and trapped? We don't know.

South Jersey's nearest match among Atlantic Coast peninsulas is the Delmarva Peninsula, and the resemblance is very close. Look at the map with your glasses off and they seem to be twin images of the same body of land. The belly of the Delmarva leads to the southwest-bound appendix of Cape Charles just as the belly of South Jersey leads to the southwest-bound appendix of Cape May. Cape Charles has a hawk count, too, at Kiptopeke, its southernmost tip. It is one of the higher counts in the East, averaging 15,000 to 20,000 hawks a year, but it has never come close to matching Cape May's 50,000 to 80,000 hawks, and those who claim to understand hawk migration have a particularly difficult time explaining Cape May's superiority to Kiptopeke. If "funnel" and "refuge" were all that made the Cape May flight what it is, Kiptopeke, not Cape May Point, would be the primary hawk-migration stop on the East Coast. Kiptopeke's east-side leading line is the same Atlantic Ocean coastline that directs hawks to Cape May; its west-side leading line is the Chesapeake Bay; and Cape Charles and the rest of the Delmarva Peninsula offer even more natural habitat than Cape May and South Jersey. Finally, Kiptopeke is directly southwest of Cape May Point! Shouldn't we see at Kiptopeke a number of hawks that have come south from Delaware, eastern Pennsylvania, and areas farther west (migrants that never reach New Jersey) *and* the hawks that departed from Cape May? Where else can all those Cape May hawks be going? Any hawk crossing Delaware Bay on a southward or southwestward track must next encounter the Chesapeake. If the funnel effect is what brings hawks to Cape May because they hesitate to cross the Delaware River, shouldn't it have an even stronger influence on the Cape Charles peninsula as the bird encounters the wider Chesapeake? And, finally, if the funnel effect is not what is primarily responsible for the Cape May hawk flight, what is?

Cape May was celebrated among birders as a migration trap long before the Cape May Bird Observatory came into existence and regular hawk counts began in 1976. Raptor researcher Bill Clark and the various teams of banders he has organized have been capturing and banding hawks at the Point since 1966; the amateur ornithologist Ernest Choate studied hawk flights here throughout the 1950s and 1960s and coordinated two season-long hawk counts in 1965 and 1970; Roger Tory Peterson, Robert Allen, James Tanner, and others conducted six season-long hawk counts at the Point in the 1930s. Witmer Stone, curator of birds at the Academy of Natural Sciences in Philadelphia, studied the birds of Cape May and South Jersey for half a century (1890–1939) and lived at the Point for the last twenty years of his life. In 1937 Stone published his two-volume classic, *Bird Studies at Old Cape May,* in which he sketched the bird life of the county, presented his theories on habitat needs and migration, and traced the records of bird observations in Cape May County back to the early 1800s when John James Audubon, George Ord, and Alexander Wilson visited. "If birds are good judges of excellent climate," Alexander Wilson wrote in about 1810, after one of his six trips to southern New Jersey, "Cape May has the finest climate in the United States, for it has the greatest variety of birds."

In fact, the history of birdwatching at Cape May goes back to at least 1633, when David Pieterson de Vries, one of the first Europeans to set foot on the peninsula, reported that the flocks of "pigeons" there (possibly passenger pigeons, now extinct) were so numerous they darkened the sky.

Nevertheless, when in the fall of 1976 an unknown young birder named Pete Dunne completed the first continuous daily count at the Point and reported a total of 48,621 hawks for the season, a widespread response among the birding community was simple disbelief. Forty-eight thousand hawks was nearly twenty thousand more hawks than Hawk Mountain had seen in its best year ever, and the conventional theory at the time was that the majority of hawks migrated along the inland ridges that led to Hawk Mountain. Migrant raptors seen along the coast were thought to be primarily straying birds that had been pushed off the main route by strong winds and had lost their way. Dunne's total was also 30,000 more hawks than the Point's own banders had predicted the count would be, and twelve times the 3,900 hawks counted by Choate in his 1965 hawk count at the Point. It even eclipsed Choate's 1970 count of 41,021—a count that had included a freakish, single-day flight of twenty-five thousand kestrels on October 16, 1970, which had been disbelieved itself.

In 1977 Dunne sat in a lifeguard chair borrowed from the state park beach patrol, and growing crowds of hawkwatchers gathered at his feet. "Birdwatchers I'd been reading about my whole life showed up," he said later. "It was all I could do to keep from climbing down and genuflecting." Some, openly skeptical of his 1976 total, came to scrutinize his identification and record-keeping skills; others wondered if the 1976 total had been an aberration—a consequence of unusually frequent northwest winds that would not be repeated. Northwest winds were even more common in 1977, however, and the count went off the scale. Dunne's final total was 81,145 hawks—an average of 146 per hour, nearly five every two minutes. "Superlatives fail under these conditions," wrote the editors of *American Birds,* the journal with the last word on such matters. "Cape May Point must now be considered the Raptor Capital of North America."

The state park built the hawkwatch platform in 1980, with room for fifty or sixty observers, and the parking lot lanes nearest it were regularly filled with the cars of birders, some of whom had driven from as far away as Kansas, Louisiana, Ontario, and New Brunswick to see the flight. In 1981 a team of three counters, sharing the season-long watch, broke Dunne's 1977 record with a count of 88,937 hawks. The next year the platform had to be doubled in size to serve the growing crowds. The deck is now thirty strides long and five across, approximately the size of two railroad cars, and should probably be doubled again. On days of northwest winds in late September and early October, when the raptor migration reaches its peak, more than a hundred observers pack together here. The platform is sometimes so crowded with visitors elbow to elbow, banging their binoculars and telescopes into each other, that the counter must retreat down the ramp to the grass below to have room to scan the sky.

The flights are still not well understood, however, and skepticism lingers about the Point's numbers. Most observers now grant that the old theory about the primacy of the inland ridge routes underestimated the importance of the coastal route, but some still believe the totals reported from the Point are too high. Visiting hawkwatchers from the inland sites tend to be especially skeptical. Counting migrants on the ridges is much simpler than it is at Cape May. To migrants heading south on the inland route, any rocky outcrop is pretty much like all the rest. They have few reasons to stop at Hawk Mountain or Raccoon Ridge or any other ridge lookout, and fewer reasons still to backtrack. You can spend a couple of days at Hawk Mountain and not see a single raptor put down in a tree; you can spend a week there and not see one reverse course. Counting along the ridges could hardly be easier: here comes a merlin; there it goes; add one to your total.

Counting at Cape May Point, on the other hand, could hardly be more difficult. The Point is very different from anything a migrant raptor has encountered before (unless he has passed through in other years), and it forces each bird to make a choice: go, delay, or stay? The ocean presses in from the left; Delaware Bay opens wide on the right; the woods and meadows below promise food and refuge; and the Delaware landscape appears low on the southwestern horizon, just close enough to offer hope for a successful crossing, just far enough to spell danger. Each hawk's decision to cross or backtrack or put down in a tree reflects a number of variables, including how far he has already flown that day, when he has last fed, how strong and from which direction the wind is coming, and what other hawks of his species are doing. As the hawks test the winds and study the

Day One

❖

other birds over the Point, their milling, circling, and reversing creates an endless series of puzzles for Cape May's census-taker. The falcons and ospreys passing on the right over the bunker and dunes, on the ocean side of the platform, sometimes swing out to sea and circle around to pass again. The broad-wings and other buteos that usually first appear to the northeast, behind the tower, often soar close to the platform and then turn back north toward the Beanery or follow Sunset Boulevard westward to drop from sight in the woods along the Bayshore. Harriers come from all sides, pumping hard and low over the wavetops on the right, gliding high at the limit of vision directly overhead, wobbling in and out of sight over the woods behind the tower. Sharpies sometimes come in overhead and fly directly out to sea, other times they join the broad-wings soaring west along Sunset Boulevard, and sometimes they stop to rest in the trees around the pond. A merlin which has been perching on the tallest tree in the merlin sticks drops from sight—chasing prey, you assume— and five minutes later a merlin pumps by on your left carrying a songbird in its talons. You didn't see it cross the pond, but merlins are stealthy and quick, so you decide it is the same bird—until you look back right and see a merlin perched in the tallest tree in the sticks again. A kettle of twenty-two broad-wings crests the horizon behind the tower, circling over the Beanery, and slowly drifts westward to drop from sight in the direction of Higbee's Beach. Half an hour later, a kettle of eighteen broad-wings appears over the Beanery, and following the same route as the earlier group, drops from sight over Higbee's. Half an hour after that, a kettle of twenty-six broad-wings executes the same maneuver. How many individual broad-wings have you seen? Were there three separate groups, sixty-six different birds? Or was it the same kettle, losing and gaining members as it circled the Point? And, what should you make of the situation on the sticks now? The tallest tree is empty, but one merlin stands on a shorter tree twenty yards to the left and another perches fifty yards to the right. What's your merlin total so far? Two, three, or four? Did you turn to watch that one with the songbird go out to sea? Or could it have swung around you and returned while you were counting the broad-wings? And, over the bunker, here comes a peregrine with a missing tail feather. Didn't you see a peregrine the day before yesterday missing that same feather? And don't some of the hawk banders believe peregrines fly out to sea for miles, then return to feed again on the shorebirds in the Meadows? And look, here's the biggest kettle of broad-wings today—fifty-five birds, low on the horizon over the Beanery. Has there been time enough for that earlier kettle to have circled all the way back from Higbee's, or are these birds an entirely different group?

First-time visitors to the Point, especially those who have learned hawkwatching on the ridges, often find the Cape May flight a disorienting experience, and they regularly question the official counter's methodology, sometimes to his face ("How do you know you didn't count that one already?"), more often behind his back ("He's double and triple counting the same birds"). More troublesome are the questions and criticisms from the local birders and platform veterans, all of whom seem to have their own pet ideas about which birds should be added to the totals and which not. Each new counter finds himself in the center of a controversy that has continued for years.

Jeff Bouton knew all this before he arrived last year. He was brought here because of the controversy that surrounded his predecessor, Frank Nicoletti. Nicoletti was finally told to quit his post, after three seasons—although he is universally regarded as the best hawk-watcher the Point has ever seen.

No one who stood with him on the platform doubted Nicoletti's ability to identify individual birds. Hawkwatching is a sport of long distances and subtle details, and Nicoletti, a burly, bearded

Day One

❖

man in his mid-twenties, saw farther and more sharply than anyone else. "He operates way out beyond me," said Pete Dunne, an extraordinary birder himself. "I think he was born with ten-power eyes." Nicoletti's dedication was also extraordinary. No single observer before him had spent 700 hours on the platform during the season, and not even the teams of observers, dividing the observation time, had totaled 1,000 hours. Nicoletti spent more than 1,000 hours each of his three seasons, and he did it alone. He arrived so often before dawn and stayed so long after dark that he began a count of migrating owls. During the day he refused to leave the platform for longer than the three or four minutes it took to get back and forth to the Park's rest room. When it rained and other birders went home, Nicoletti sat in his truck in the parking lot with the engine running and wipers going, on the off-chance that a hawk might fly by. Two months after the fall count ended in December, he began another count, a February-to-June census of northbound hawks at Braddock Bay, New York, on the shore of Lake Ontario. When the Braddock count finished, he worked as a short-order cook for a few weeks in July, then reported back to the Point on August 1 to start again. "The Iron Man," he was called, and by late October his eyes were so red from scanning the sky they glowed like a pair of brake lights.

But all of Cape May's census-takers before Nicoletti had used mechanical hand counters to keep track of their totals. On days of big flights most had used three hand counters simultaneously—one for sharp-shins, one for kestrels, and the third for whatever species happened to be the next most common hawk of the day. Nicoletti refused to use hand counters at all, insisting he could keep all counts in his head. He also refused to total the birds at regular intervals. Even on slow days census-takers are supposed to note totals every hour on the hour—along with wind, temperature, and cloud-cover data. Nicoletti often seemed to let several hours pass before he picked up his clipboard to pencil in his numbers, and the daily totals board hanging from the platform seemed always to be at least three or four days out of date.

Nicoletti's counts for merlin and peregrine falcon were especially high. For the eight years of counts before Nicoletti's tenure, the average merlin total was about 1,000 birds a season; for Nicoletti's three years the average was more than 2,500. In his first year on the platform, Nicoletti's count for peregrine was 518, two hundred more than had ever been seen there in a single season. In his third season he reported 615 peregrines, a number higher than most estimates for the entire eastern North American population of the species. Were the populations of merlin and peregrine exploding? Or were Nicoletti's totals careless guesstimates?

At least as troublesome for the Cape May Bird Observatory (CMBO) as questions about these numbers was Nicoletti's demeanor on the platform. He seemed aloof and rarely spoke to any visitors except a handful of platform regulars. Other counters had thrived as the natural center of attention, using the position as a chance to interest visitors, most of whom are novices, in hawks and hawkwatching. Nicoletti preferred to stand silently in the least crowded corner of the platform, sweeping his binoculars back and forth across the sky.

His defenders argued that Nicoletti's style was a consequence of his intensity—"Frank wants to identify every hawk first"—and that his counts were higher than those before him because he saw farther and paid more careful attention than other observers.

Two years ago, when CMBO was looking for a counter for the spring hawkwatch it sponsors in Sandy Hook, on the north shore of New Jersey, Nicoletti recommended Jeff Bouton, a young birder he'd met at Braddock Bay, and Bouton took the job.

In a three-month census at "The Hook," Bouton proved to

be a skilled observer who was careful about keeping his data sheets complete and up-to-date and talked with anyone who happened by. Late that summer CMBO called him again. Frank Nicoletti would not be allowed to continue as CMBO's official counter in the fall. Would Bouton like to replace him?

The offer put Bouton in an awkward position. "How could anyone who loves raptors turn down a chance to be the counter at Cape May Point?" Bouton remembers thinking at the time. "But I figured I had to say no. Frank was my mentor. Then he called me up. He'd heard I'd been offered the job, and he told me to take it. He said he was pissed about getting pushed out, but he'd be even more pissed if someone he didn't know took his place."

Bouton counted until December 6 last year, and his totals for the season, written in grease pencil eight months ago, remain on the totals board nailed to the side of the platform.

It was a mediocre year by the Point's standards, except for the osprey and Cooper's hawk counts, both all-time records, and the counts for the two species whose totals had seemed most out of line during Nicoletti's years, the merlin and the peregrine. There were no complaints about Bouton's censusing techniques, but his merlin count was right at Nicoletti's average, and his peregrine count was seventy birds *higher* than Nicoletti's highest count.

For the low counts on the other birds, Bouton wonders if he is partially to blame. "I'm sure I missed some Swainson's hawks," says Bouton. "The first one I saw Frank had to show me. He came walking over one day, not even wearing binoculars, and pointed up in the sky. 'Hey, there's a Swainson's!'

"I think I might have been too conservative counting eagles, too. I don't like to count them until I see them fly over the horizon and out of here, but I wonder if I undercounted them. Everybody warned me the first season would be a learning process, and they were right.

Cape May Point Hawkwatch

SPECIES	YESTERDAY'S TOTAL	TOTAL TO DATE	PEAK FLIGHT AND DATE
TURKEY VULTURE	10	530	53 10/22
GOSHAWK	1	24	3 11/5 & 11/9
COOPER'S HAWK	1	2,950	291 10/13
SHARP-SHINNED HAWK	8	19,886	1,131 9/25
RED-TAILED HAWK	54	1,868	296 11/22
RED-SHOULDERED HAWK	—	651	70 10/31
BROAD-WINGED HAWK	—	4,608	1,400 10/5
ROUGH-LEGGED HAWK	—	4	—
SWAINSON'S HAWK	—	3	—
GOLDEN EAGLE	—	11	3 10/25
BALD EAGLE	—	49	5 10/9
NORTHERN HARRIER	26	3,080	187 10/4
OSPREY	—	5,402	508 9/21
PEREGRINE FALCON	—	686	116 10/17
MERLIN	—	2,447	259 10/1
AMERICAN KESTREL	—	16,532	3,694 10/1
BLACK VULTURE	3	3	3 12/6
TOTAL	103	58,734	

"One day I turned around and there was a beautiful blond sitting on that bench right there. Really dynamite. Best looking woman up here all season. The platform was crowded, and I tried to play it cool—making my way toward her real slow, identifying birds for people along the way. But I stopped to show someone an osprey, and next thing I looked up and she was getting into her car out in the parking lot. *Alone.*" He shakes his head, "That was the worst mistake I made all year. Young female birdwatchers are extremely rare."

By 10:00 A.M. today Bouton has recorded eighteen hawks: six harriers, nine kestrels, and three broad-winged hawks. "For August this is a good flight. Last year I had zero the first two days and one solo red-tail the third day at two-thirty in the afternoon. By then I was about ready to pack up the shop for the season." He pivots and follows other birds with his binoculars: a royal tern winging up the dunes, a white-rumped sandpiper trailing a flock of semipalmated sandpipers, a rough-winged swallow twirling by—"All right, rough-wing. All I need now is cliff swallow, and I've got all five swallows the first morning."

The first visitor to come up the ramp is a man from Pennsylvania. "Where's the hawk counter?" he asks Bouton.

"That's me."

"What happened to that heavy-set guy with the beard and the red eyes?"

"You mean Frank Nicoletti. He worked in the banding stations last year. This year he's going to Israel for the migration there."

"I've never seen anyone who could identify hawks like him."

Bouton nods, "I'll tell you what I told everybody who said that last fall. I'm not Nicoletti. I'm going to make some mistakes. But we'll live through them together."

The Pennsylvanian works in Harrisburg, and his office faces the Susquehanna River. "I keep a scope on my desk aiming out the window. This summer I had an osprey fishing on the river almost every day. In the winter I've always got ducks and cormorants."

"Two more kestrels," says Bouton, pointing. "Just right of the second insulator on the right-hand guy wire."

"This is remarkable. It's too early for these birds to be migrating."

"Actually, kestrels are usually moving by this time of year," says Bouton. "They're early migrants."

"I thought broad-wings were the early migrants."

"Kestrels and broad-wings both move early. And harriers. Harriers migrate every month and in every kind of weather. Rain, snow—they don't care. Even at night. I've been out here after sunset and seen harriers pumping down the beach heading south."

"Here's a harrier right now," says Bouton, a moment later. "Adult female. Dead ahead in the gap."

The Pennsylvanian steps behind Bouton's shoulder and swings his binoculars into line. "You can get color off that bird already?"

"Not much. It's a female because the wings are so broad. Males have thinner wings. And it's an adult because it's molting. Immature harriers never molt in the fall. See the missing feathers when it banks? . . . *There.*"

"Amazing," says the visitor. "Amazing. I could learn a lot if I came here for a week or two. Isn't this a fascinating hobby?"

"Hobby?"

"Well, I guess with you it's a business."

Bouton takes a breath. "Actually, it's going to have to become a hobby. One of these days I'm going to have to fix my car, which means I'll need a real job. Which means I've got to go back to college."

"How much do they pay you for doing this?"

"Not enough to fix my car. Or go to college. But I figure in the meantime I'm getting paid to do what I'd otherwise do for nothing."

Bouton is paid $1,500 for his three-month census, about $1.60

per hour. He spent last winter working as a waiter in Rochester, New York, where he grew up. Like Nicoletti and several others who have served as CMBO's hawk counter, Bouton is a self-taught naturalist who found classrooms too confining to bear. "I went to college for a year, then I left. I learned more on this platform in four months last fall than you could possibly learn in four years of school. There's always something new to see here.

"Variety," says Bouton, changing the subject. "That's what we have that Hawk Mountain doesn't have. They have the pretty views on those ridge sites. I'll give them that. But last year two hundred and thirty-seven different species of birds flew over my head on this platform. I like to call up Laurie Goodrich, the counter at Hawk Mountain: 'Hi, Laurie. How many sandhill cranes have you seen lately?' "

Ten minutes after this comment, Bouton points at a bird zipping over the pavilion to the left of the platform. "Hey! Look at that rump patch. Cliff swallow!"

A woman in her sixties, holding a pair of binoculars hardly bigger than a deck of playing cards, walks up. "Can you tell me what that bird is, hovering over the water there?"

"That's a kingfisher," says Bouton. "See the crest and the blue back?"

"My goodness. We certainly don't have those in Connecticut."

"Oh, I think you do if you look for them."

The woman aims her binoculars at the horizon. "What's this big white bird over here?"

Bouton steps behind her and sights over her head. "If I'm looking at what you're looking at, it's a plane."

"Goodness gracious! I'm sorry."

"Nothing to worry about," says Bouton. "I once tried to turn a monarch butterfly into a hawk."

The hawkwatch
platform

A tall birder walks up the ramp. He is from Kent County, England. He nods at the kingfisher still hovering over the pond. "There's a lovely bird," he says. "We had one in the Scilly Isles last fall. That's our Cape May, our best migrant trap; lots of stuff fetches up there—North American vagrants and other goodies."

He points to the house sparrows under the pavilion. "Now there's a bird that is altogether *too* familiar. What I don't understand

Day One

❖

is why you Americans wanted a pest species like that over here." The house sparrow is native to Europe and Africa and was intentionally introduced in the United States in the mid-nineteenth century. Today it is one of the most numerous and least loved birds in North America.

"Oh, they serve their purpose," says Bouton. "Food for raptors."

"Yes, I suppose that's right. Everything finds its place."

"I like to tell people there are only two kinds of birds we see from the platform: hawks and food for hawks. When the dicky birders come charging over here all fired up about some rare warbler they just found out in the woods, I like to tell them, 'Ah, warblers are just hawk bait.' That always drives them crazy. They hate it when they see a raptor nab a songbird. I *love* it. Last year I saw thirteen kills by peregrines. One afternoon two immature female peregrines worked a pigeon over the bunker, chasing it back and forth between them until one of them finally nabbed it. Then the next day two female peregrines, probably the same birds, caught a goldfinch in the air with the same kind of teamwork. Where else but Cape May can you see something like that?"

At 11:40 A.M. a trio of turkey vultures pushes the count over thirty raptors for the day. Meanwhile, the weatherman on Bouton's radio announces that a cold front is passing the Great Lakes tonight and will reach the Delaware Bay area sometime tomorrow afternoon.

Bouton pumps his fist like a football fan anticipating a touchdown. "All *right*. Passing the Lakes tonight, and it will be here tomorrow? That front is *moving*."

Hawkwatchers are empiricists, by and large. Ask the veterans on the platform "What brings the hawks to Cape May?" and you'll likely not hear a word about the Point's peninsular geography or its mix of natural habitats. The "funnel" is too large to see, and the "refuge" is taken for granted.

"Cold fronts," they'll tell you. "Cold fronts and northwest winds."

Spend two weeks on the platform and you'll likely adopt this abbreviated understanding for your own—and find yourself on the edge of your chair each night as the TV weatherman cues up tomorrow's weather map. Hawkwatching is always good at the Point after a cold front passes, and it is best when winds have been strong and northwesterly. Under other conditions—"back door highs" and southerly winds are the worst—hawks can seem depressingly scarce in the Cape May sky.

Ask the veterans on the platform for a longer analysis of the cause-and-effect relationship between raptors and weather systems, however, and the most frequent answer you'll get is a shrug and a smile. Where do all these hawks come from? "Somewhere north of here." Where will they be tomorrow? "Who knows?"

The most widely accepted theory connecting the Cape May hawk flight and northwest winds is the so-called "wind-drift hypothesis." It was first put into print more than fifty years ago, in 1936, by Robert Allen and Roger Tory Peterson and no longer fits all the evidence, but it has been repeated in slightly different versions in dozens of articles and books and is still the most coherent explanation of why Cape May sees more hawks than other coastal peninsulas. The source of Cape May's hawks, says wind-drift theory, is in the opposite end of New Jersey—in the Kittatinny Mountains in the northwestern corner of the state. The Kittatinnies are the eastern ridge of the Appalachian Mountains, which are still believed to be the preferred route for most species, and which form, it would seem, an ideal pathway for southbound raptors, running from New England to Georgia. Under most weather conditions, updrafts blow off the faces of the ridges and give migrating hawks extra lift, saving them energy on their long flights. Cold fronts, which trigger migration by dropping local temperatures and clearing the skies, generally blow diagonally across the ridges, accompanied by northwest winds. Wind drift occurs when the winds blow so strongly that hawks are pushed off the ridges and driven down into the flatlands of central New Jersey. From there, the funnel effect leads them to the Point.

Cape May's banding records support the wind-drift hypothesis. More than seventy-five thousand hawks have been banded at the Point since 1966, yet only a handful have ever been recovered in South Jersey in the seasons following their original capture. Most recoveries in subsequent falls have come from inland areas, the ridges and elsewhere. That suggests that the hawks who are funneled through Cape May learn it is a dead end and do not repeat their mistake in later migrations. The disproportionate number of immature hawks seen at the Point also supports this interpretation. Immatures outnumber adults in all hawk species at the Point. Both of the common accipiters, the sharp-shinned and Cooper's, can be separated into their two age groups at long distance—adults have rufous breasts and blue-gray backs; immatures have grayer breasts and plain brown backs—and the imbalance between adults and immatures is extreme. Ninety-five percent of the accipiters at the Point, nineteen of every twenty, are immature birds, making their first migration. On the ridge lookouts, adult accipiters are generally at least as numerous as immatures and often outnumber them. First-year accipiters are very young, only three or four months old, when they begin their migration, and they are inexperienced flyers. Since accipiters are deep-forest birds on their nesting grounds, who live by chasing songbirds and other prey on short, dashing flights from one perch or another, they are particularly unaccustomed to battling contrary winds over long distance as they must during migration. First-year migrant hawks also have only instinct to direct them on their path south—they have never been where they are going, and they have left their parents behind. We would expect immatures to get lost more often than adults, so Cape May's ratios seem to support the idea that the birds seen there have drifted from the track they should be following. If the coast is not the inferior and more dangerous route, why do so few adults follow it?

Some observers have suggested that in the different proportions of immatures and adults on the ridges and at Cape May we can see selective pressure in action. Those few immatures who battle the contrary winds and hold to the ridge route are the birds more likely to complete their migration successfully and return north in spring to breed. Adults on the coastal route are few because most of those who try to follow it south die in their first attempt.

But cold fronts and northwest winds cannot be having as direct an effect on Cape May's migration as the wind-drift hypothesis makes it sound. For one thing, the connection between northwest winds and hawk numbers is much more localized than the theory suggests—as any disappointed hawkwatcher can attest who has raced through a storm from his home in central or northern New Jersey and discovered a blue sky and no hawks at Cape May Point. It is northwest winds *on the peninsula itself,* not those in upstate New Jersey, that create the best flights at the Point. Cold fronts that stall or blow out to sea north of the peninsula have little effect on the flights at the Point.

Another complication involves the sheer numbers of hawks at the Point. Can so many hawks be off course every year? Can fifty to eighty thousand hawks be wrong? In autumn 1984, watchers on the ridge lookouts were disappointed as week after week, month after month, the whole migration season passed without any significant cold fronts. "The Fall with No Fronts" it came to be called. Totals for all species were down at most inland lookouts, and crushingly low at several. "1984 will soon be forgotten," wrote the compiler of the Hook Mountain (New York) count at the end of the season, "and I'm glad." 1984 won't soon be forgotten at Cape May. The count was 83,548 raptors, the second highest total ever recorded there, including 61,180 sharp-shins, the highest total for any species ever. If it's northwest winds that blow sharp-shins off their preferred path, why was the record set in a year without fronts? And why were there so many more sharpies at the Point that fall than on the ridges? The combined total of sharp-shins recorded at all twenty-four ridge lookouts in New York, New Jersey, and Pennsylvania was 29,197, half the number seen at New Jersey's single coastal lookout. Could 61,180 sharp-shins have been on the wrong path—in a fall without fronts?

Selective pressures must be at work during migration. Migrants of all ages and all species of birds die each fall trying to reach their wintering grounds, and the first migration is the most dangerous of all. Far fewer individuals of all species fly north in spring than started south in fall. But it cannot be that the ridges are so much the better route and the coastal route so inferior that the ridge migrants are winning the race and the Point's migrants heading toward their doom. If that were so, those who lacked the instinct to follow the ridge route would have been eliminated from the gene pool long ago. If the coastal route were so bad, selective pressure and evolution would never have allowed it to develop into what it is—the primary route south for tens of thousands of immature hawks. Cape May's consistently high counts tell us something positive is gained (or something negative avoided) by the hawks that take the coastal route; we just don't know yet what that something is.

Finally, one last complication with the wind-drift hypothesis leads to the most intriguing idea of all. The complication was first suggested by Witmer Stone in 1937, just one year after Allen and

Peterson published the original version of the wind-drift theory, but it has been largely ignored until recent years. Stone, an extraordinarily diligent observer of birds, noticed fifty years ago what birders with far better optical equipment still miss today: hawk flights at the Point on the "wrong" winds. In his *Bird Studies at Old Cape May,* he reported, "With winds unfavorable for the usual flights [*i.e.,* from east or south] I have often seen, with the aid of binocular glasses, large numbers of hawks circling high overhead, so high indeed that they appeared like small swallows or even insects, and gradually drifting off to the south." The northwest winds might not create the hawk flight, he concluded—only bring them low enough to be seen, as the birds fight against being blown out to sea. On other winds, they fly so high we miss them. In other words, the connection between northwest winds and migrant hawks might only be an optical illusion! "If this is so," Stone concluded in a wonderful understatement, "it follows that the hawks counted during a flight are only a small proportion of those actually on migration."

In 1982, forty-five years after Stone's book was published, two ornithologists drove to the Point in Clemson University's Avian Migration Mobile Research Laboratory, a motor home equipped with a marine surveillance radar, a fixed-beam vertical radar, night vision devices, and video recording equipment, to test his hypothesis. After a month of observations, they wrote an article for *Animal Behavior* supporting Stone and others who believed the wind-drift theory was too simplistic. Hawks seem more numerous on northwest winds, their instruments indicated, only because they fly much lower in those conditions. They fly so high on east winds, averaging 900 to 2,600 feet up, that most cannot be seen even through binoculars without cueing by radar—unless the birds are directly overhead and so in full silhouette. Ospreys, turkey vultures, and sharp-shins were identified by radar and telescope at 3,200 feet, beyond the limits of binocular vision, and one unidentifiable bird was detected at 3,800 feet.

If Stone and the Clemson researchers are correct, Cape May's count is not too high, as the Point's critics maintain. It's far too low. The actual number of raptors migrating over the Point may be three or four times the number counted from the platform. Perhaps a quarter of a million hawks pass through each fall. And those who want to explain the Point's flight have an even larger puzzle on their hands.

At 2:00 P.M., for the first time all day, Bouton has no hawks to record in his hourly compilation. The wind at the platform is now light and southerly, although out beyond Cape May City a line of dark clouds seems to be moving along the ocean's horizon from the north. "That looks promising," he says. "A little prefrontal action."

Richard Crossley arrives on his bicycle, a rented, rusting three-speed with wire baskets over the rear wheels. Crossley is a twenty-four-year-old Englishman who looks like a young Kirk Douglas, with a dimpled chin, green eyes, and sun-lightened brown hair. He wears sandals, Bermuda shorts, and an unbuttoned shirt. Only his well-worn binoculars—one eye cup torn, both barrels rubbed smooth from long use—indicate he is a birder. He and David Sibley, an American, have been conducting a songbird survey for CMBO at Higbee's Beach, counting migrating warblers, vireos, and other passerines each morning since mid-July. At night Crossley works as a waiter in a Cape May restaurant.

"Yesterday morning at Higbee's was fantastic," he tells Bouton. "Fantastic. Eleven hundred redstarts, a hundred and sixty northern waterthrushes, three mourning warblers, two ceruleans. And we didn't expect anything. I was out drinking until one A.M. the night before, and when I stepped out of the bar I heard the chips of the

Sharp-shinned hawk
flying past the old Coast
Guard tower

birds going over. Hundreds of them. That's when I knew I would have to be on the dike at dawn, three hours later." He sticks out his tongue and wobbles his head like a clown drunk. "Maybe I was seeing double. Maybe there were only five hundred and fifty redstarts."

The sky to the east darkens around 3:00 P.M., a cold wind begins to blow, and thunder and lightning follow. The parking lot clears quickly as beachgoers hurry to their cars. No hawk has passed for a couple of hours, but Crossley and Bouton remain on the platform, pointing at the lightning bolts over the city and the ocean. The poplars across the pond on the edge of the Meadows flash the silver undersides of their leaves as the branches bend, then they blacken as a huge, crescent-shaped cloud sweeps overhead. Crossley studies it through his binoculars. "Sometimes the birds fly ahead of storm clouds."

"This must be the leading edge of that Great Lakes front," Bouton says. "It's really coming strong. I love it."

"Too northeasterly," Crossley says, shaking his head. "What you want is something with a westerly component. That's what brings the warblers. That's when they get blown out to sea and have to fight back this way."

"Last year," says Bouton, "it seemed that when the front came through determined how many warblers we had the next day. The warblers were always on the back side of the front. If a front didn't come through before midnight, we got no songbirds the next day."

Crossley agrees. "Warblers need the stars to be out or they won't fly, so the sky has to clear early enough in the evening for the birds to take off. If it's a cloudy night, they just stay where they are.

"You see how much more interesting songbirds are than raptors?" Crossley continues. "Their migration is still pretty much a blank—a bloody unknown *blank*. Nobody's investigating it, I mean really investigating it. I'd like to get some of those funds devoted to you and the hawks diverted to the passerines. Why the hell should all that money go to hawks?"

"Yeah," says Bouton, "all that money."

"Seriously, why is it you Americans like hawks so much? It's because they're so big and slow and easy to identify, isn't it? The only other birds Americans really care about are ducks. Which are the only birds even slower and easier to identify."

As raindrops splatter on the deck, Bouton and Crossley retreat under the pavilion next to the platform and sit at a picnic table. Crossley reaches for Bouton's scope to follow a tern flying over the dunes. "Let's just make sure it isn't something it's not supposed to be."

A middle-aged, shirtless jogger trots up, dripping wet. He is wearing a Walkman around his ears, which he peels off as he approaches the pavilion. "Hi, Jeff!"

"Hey, George."

"Welcome back to Cape May."

Bouton stands and shakes hands. "Thank you."

"You doing the counting thing again?"

"One more time."

"How is the weather we've been having going to affect the hawks? It's been so hot this summer—the Greenhouse Effect and all. Are we going to get as many birds as we usually do?"

Bouton shrugs, then raises his binoculars to follow Crossley's tern, which has looped over the ocean and is headed back up the line of dunes. "I don't know. I'll tell you what. Ask me again at the end of the season, and I'll give you my answer."

Chapter Two

❖

BEGINNINGS

ONE MARCH EVENING IN 1974, CLAY SUTTON WENT TO A MEETING OF the Izaak Walton League, a hunters and fishermen's conservation group with a chapter in Cape May, carrying a photograph of a kestrel to show to a local naturalist named Al Nicholson. Nicholson, a tall, straight-backed man in his late forties, was said to know the locations of the nests of the last two pairs of bald eagles in New Jersey. Sutton had long wanted to talk to him. The photograph was an offering.

Sutton considered himself neither a photographer nor a birder, though birds had been tugging at his consciousness all his life. He had grown up in Stone Harbor, New Jersey, a few blocks from the ocean, and cormorants, gulls, and herons were more common than squirrels in his neighborhood. His father and grandfather hunted ducks and quail, and his grandfather carved duck and shorebird decoys. Once each summer, in a boyhood ritual he continued into his teens, Clay had hiked the eight miles back and forth to the tern and skimmer colony then at Stone Harbor Point to watch the birds brooding their eggs and feeding their young. But he had no birding mentor, and when his grandmother gave him a fifty-cent guide to North American birds when Clay was nine or ten, the golden eagles, black-throated blue warblers, and other birds pictured there seemed as mysterious as creatures in a science fiction fantasy. How could anyone hope to see all these strange birds in the real world? he'd wondered. Coming home from school one autumn day, he spotted an odd little bird walking from branch to branch in a bush in the front yard. He went up to his room and was startled to find its picture in his book: it was a black-and-white warbler.

Migration intrigued him. Surf fishing in high school he noticed that ospreys and other birds seemed to come south at the same time the fish did: in September and October, after cold fronts and northwest winds. One day, casting from the beach, he watched a songbird flutter in from the ocean. It landed on the tip of his pole and perched there, peering down at him, first with one eye then the other. Another time, a tiny bird flew in off the ocean so weak that it collapsed and fell in the surf a hundred feet from land. Sutton waded out, scooped it up, and carried it to the dunes. Half an hour later it recovered, shook itself off, and flew up again, continuing on its way south. Sutton didn't know what kind of bird it was, couldn't find it in his book, and didn't know anyone who might identify it for him.

He majored in biology at Gettysburg College, in Pennsylvania, imagining he would study wildlife, but the curriculum was intended for premed students. It wasn't until the summer of 1972, a year after he graduated, when he accompanied one of his professors on an expedition to study bats in Chiapas, Mexico, that Sutton got his first taste of field biology. The students trapped the bats each night and in the morning, after they took down the nets, the professor led a walk through the forest. He showed them keel-billed toucans, barred antshrikes, collared aracaris, and dozens of other tropical birds.

One evening, walking alone on a beach near Tonalá, Sutton spotted a beautiful white bird framed against the sunset while it coursed the dunes. His field guide told him it was a raptor, a white-tailed kite, and that the species nested in the United States. Something stirred inside Sutton that moment.

Home again in New Jersey the following fall, Sutton completed a Master's degree in environmental education, then found what he assumed would be a temporary job as a lab technician with the Cape May County Health Department. He had taken up surf fishing again, and began wondering how he might see an eagle. Two pairs of eagles still nested in southern New Jersey, he had heard, in Bear Swamp near Dividing Creek on the upper Bayshore. One of his fishing friends had told him that Al Nicholson knew where they were. Nicholson was a local celebrity, regularly interviewed in the newspapers because of his opposition to the County Mosquito Commission and other government agencies he believed threatened the environment. Sutton had seen him at Izaak Walton League meetings, but had never spoken to him.

One late winter day Sutton drove up the Garden State Parkway to Brigantine National Wildlife Refuge and rode around the dikes. A kestrel hovered over the marsh, beating its wings back and forth. Sutton pulled over within ten yards, reached into the back seat for his camera, leaned out the window, and snapped the shutter. It was the first hawk photograph he had ever taken, and it is still, ten thousand photographs later, one of his best. The bird was an adult male in fresh breeding plumage, and the camera caught it with tail flaring and wings stretching to their fullest extent. The blue sky above is cloudless. The bird is looking forward, head held high, as if contemplating a long flight home. The black mustache mark descends down the cheek like a battle scar. The wing linings are translucent and as neatly contrasting as a checkerboard.

This was the photograph Sutton brought to the meeting to

show Nicholson. Nicholson studied it for a long time—he was an artist and a connoisseur of photographs, especially raptor photographs. "I ought to take you with me next time I go up to Dividing Creek," he said at last. "We might see an eagle or two."

The following weekend Sutton found himself trailing Nicholson through a labyrinth of sand roads west of Bear Swamp. Nicholson wore heavy rubber boots, carried a huge pair of binoculars hanging from a rawhide string, and chain-smoked unfiltered Camels, but the pace was hard. Nicholson had a pounding, fast stride, and he paused only long enough to light his next cigarette. Within the first hour Sutton noticed they'd passed the same tree three times and realized they were walking in circles. Nicholson was trying to confuse him, he guessed, or testing him, making certain Sutton earned his eagle. After another hour's marching, Nicholson suddenly turned down a path they had passed twice earlier, walked a quarter-mile in, and pointed. An adult bald eagle, huge black wings and bright white head, soared over them. A few moments later a second adult flapped into sight and joined it in the sky. Soon the two birds were calling to each other in their high-pitched voices and exchanging sticks in the air. "Now," said Nicholson, "doesn't that just rejuvenate your soul?"

Sutton had found his mentor.

Nicholson had been studying raptors at the Point since the 1930s, and he seemed a man from an even earlier age. He was a landscape painter who worked with oil on canvas, always outdoors under natural light. His favorite theme was the play of sun and cloud. When the weather was good, he painted outdoors as often as possible from May through November at various spots around the peninsula. On rainy days and evenings he pored through collections of nineteenth-century photographs and studied the clouds and trees in the background. "Look at that," he'd say, tapping a landscape from

Miller's *Photographic History of the Civil War.* "Look how the light is striking that maple. That's really interesting." His most treasured book was *Bird Studies at Old Cape May,* Witmer Stone's two-volume natural history of the South Jersey peninsula. Stone's eloquent essays on heronries and sparrow colonies described the Cape May Nicholson had loved best. The black-and-white photographs that accompanied the text—crystal-bright pictures of spongy and untrampled ponds around the Cape May Lighthouse, wide and empty salt meadows at Seven Mile Beach, and gnarled trees shouldering sand in the Bayshore dune forest—showed the Cape May that had been lost. Nicholson spoke frequently of places Stone had known—Bay Shore Meadow, Two Mile Beach, Price's Salt Pond—areas whose names only the oldest local residents now recognized because they had been changed beyond recognition or had disappeared entirely.

Away from his easel, Nicholson spent most waking hours, sometimes forty or fifty hours a week, battling and raging against countless adversaries: the banders who trapped hawks at the Point each fall, the joyriders who raced jeeps and four-wheel-drive trucks through New Jersey's last remnant of dune forest at Higbee's Beach, the County Mosquito Commission whose helicopters sprayed insecticides and whose ditches drained marshes throughout the peninsula, the County Planning Board which year after year allowed tracts of woods to be replaced by malls and shopping centers, the County Freeholders who refused to restrain the Mosquito Commission or the Planning Board, the State of New Jersey which in 1954 had extended the Garden State Parkway into Cape May County and so encouraged destruction by increased population pressure, and the City of Cape May which in 1942 had constructed the Cape May Canal and so cut off the City and the Point from the natural life of the rest of the peninsula like a man sawing off the tree limb he sat upon. Nicholson was tireless. He diligently attended every Mosquito Commission meeting, every Planning Board meeting, every Freeholders' meeting, and then returned to his house at night to write out fiery letters and position papers for CAPE (Citizens Association for the Protection of the Environment), the local conservation group he had helped form. His anger was fueled by a despair that the best of the peninsula's wilderness areas were already gone, and his basic political technique was full-throated confrontation. "You people," he shouted at one meeting, finger pointed at the County Freeholders, "don't deserve to live!"

Once a week or so, Nicholson would retreat into the woods. "Let's go really deep into Buckshutem Swamp," he'd say to Sutton, "or let's hike across the Pine Barrens." He was an indefatigable walker with a long stride, and though he always carried his binoculars, he bristled at any suggestion that the purpose of these outings was merely birdwatching. He had his eye out also for moving clouds, swirling water, cattail ponds, and especially old trees. Coming upon a three-hundred-year-old oak deep in the forest, he would cry out, "Look at that! Doesn't that rejuvenate your soul?" Some trees he knew individually, "Classical Trees" he called them—"Let's go look at the Classical Tree at the Bevin Tract," he'd say to Sutton, and "Let's go visit the Seven Wonders," referring to a certain cluster of seven huge tulip trees deep in Bear Swamp.

Birdwatching was secondary, in Nicholson's mind, to appreciating "the conditions," the ebb and flow of sun and landscape. Birds were only part of the whole, elements in an ever-changing composition. A passerby, seeing Nicholson's binoculars, asked him one day, "What are you looking for?" Nicholson pulled himself up and studied the man. Then, with exaggerated enunciation, like a full professor addressing a freshman, he answered: "Colors . . . tones . . . light . . . and *life!*"

Still, Nicholson was fascinated with birds, raptors especially.

"I had no idea you could see hawks in numbers like that . . . "

He slogged through Bear Swamp in hip boots each spring, searching out hawk and owl nests, and he spent dozens of hours each fall studying hawk flights from the Meadows, the Beanery, Town Bank, and other lookouts around the peninsula. One day in September 1974, Nicholson took Sutton with him to the Point. The winds were from the northwest, and hawks peppered the sky: kestrels, sharp-shins, ospreys, broad-wings. They found three hundred broad-wings circling over New England Road. Sutton was stunned. "I had grown up only fifteen miles away. I had a vague idea about Cape May being a place where birders went to watch migration. But I had no idea you could see hawks in numbers like that anywhere in New Jersey."

* * *

Remarkably few people knew about the Cape May hawk flight before 1976, the year Pete Dunne conducted his first season-long count. How much credit Dunne should be given for the Point's current reputation and for the thousands of birders who now visit each fall is a matter of debate, one that usually splits along generational lines. Dunne's admirers tend to be birders his age or younger. When they refer to the years before 1976 as "prehistory" or "B.P." (Before Pete), they infuriate those who have been birdwatching here since the 1940s, '50s, or '60s. Older birders know Cape May's ornithological history is as long and rich as that of any place in North America, and they remember fondly fall weekends when birds were even more numerous than they are today and birdwatchers much scarcer. The most veteran observers generally believe that Dunne had little to do with the crowds and other changes that have come. The Point's new celebrity is, they say, the inevitable result of the increased interest in birds, especially in hawks, that has been a nationwide phenomenon in the last two decades.

No one argues that the scene at the Point today is anything like it was before 1976. In the "B.P." years the birders who regularly visited were so few that most knew each other by name, and a cluster of five or six people wearing binoculars would draw stares from passersby. Nicholson and Sutton could spend a whole day watching hawks from the Beanery or Higbee's Beach and encounter no other watchers. Even at the Lighthouse they might find only one or two other birders. Climb the Lighthouse on a weekend in September or October today and you can count two or three hundred birders below you—on the platform, around the parking lot, and along the edge of the park's woods. Drive over to Higbee's Beach on a Saturday morning after a Friday night cold front and you'll find the line of birders' cars parked along New England Boulevard stretching for half a mile. Fifteen years ago Higbee's only claim to fame was its reputation as the most accessible nude beach in New Jersey. Today naked sunbathers are rarer there than western kingbirds.

Pete Dunne's arrival at the Point in 1976 was not the sole factor that led to these changes, but it was certainly an important one. Hawk Mountain has been drawing thousands of watchers every fall since the mid-1950s, many of whom were coming from Philadelphia and points east that were actually closer to Cape May. The old misconception that the majority of hawks migrated along the ridges rather than the coast had something to do with this; more crucial were the contributions of the two people who had pushed Hawk Mountain Sanctuary into the limelight. The first was Rosalie Edge, the founder of the sanctuary, a fiery, articulate, and hard-driving conservationist who realized that the best way to protect the hawks on one ridge in Pennsylvania was to publicize the magnitude of the hawks' flight and their slaughter by local hunters to people who had never visited the area. She single-handedly raised the funds to purchase the sanctuary's land in 1938, served as the sanctuary's president for more than thirty years, and continued to champion the conservation of birds of prey beyond the sanctuary's borders until her death in 1962. The other crucial person in Hawk Mountain's history was the man Mrs. Edge hired as the sanctuary's curator: Maurice Broun. Broun was a self-educated biologist, a brilliant observer of the natural world, and a masterful teacher. He lived at the sanctuary for more than forty years, patrolling its borders, counting the migrants, charting the winds, developing new identification techniques, leading countless field trips, and speaking to visitors at all levels of knowledge, from Cub Scouts to foreign ornithologists. At first Broun's unofficial title was "Keeper of the Gate." As more and more people came to meet him and learn from him, and as Hawk Mountain's reputation grew so large that it seemed a kind of Olympus to many visitors, Broun's unofficial title changed. Soon he was Hawk Mountain's "Keeper of the Flame."

The Point had neither an Edge nor a Broun. Cape May's two leading hawkwatchers in the 1950s and 1960s were Al Nicholson and Ernie Choate, an English teacher and author of *The Dictionary of American Bird Names.* Choate lived at the Point and shared Maurice Broun's interest in careful, long-term record-keeping—he had compiled the county's bird records for years, had updated Stone's tables for the 1965 republication of *Bird Studies at Old Cape May,* and had conducted two season-long hawk counts in 1965 and 1970—but he was neither the naturalist nor the inspiring personality that Broun was.

Al Nicholson shared Rosalie Edge's fighting spirit, but he lacked her interest in drawing new converts to the cause. In many ways Nicholson operated as an antipublicist. "Every election year we get back in touch with mainstream America," he once told a journalist. "It's a *horrifying* experience." Nicholson wanted privacy on his tramps through the woods, and he didn't like birders. "Seen anything really interesting today?" he liked to ask those few he encountered. "No," they'd answer. "Well," he'd snap, *"you* probably won't." Birders weren't fighters, Nicholson believed, they were recreationists. In environmental confrontations, when the time came to attend meetings, to write letters, to stand up to developers and politicians, birders were off chasing warblers. They wanted to enjoy the natural world, Nicholson believed, without investing any energy in defending it.

Clay Sutton accepted these opinions as readily as he accepted Nicholson's advice about where to look for Cooper's hawks or about what winds led golden eagles to Pond Creek. Sutton knew no other birders in these first years and made no attempt to meet any. Knowing Nicholson was enough. The man was a storehouse of raptor lore, much of it so intuitive he couldn't explain it to Sutton, only demonstrate it. "Look at that lambent light," he'd announce, "the eagle is imminent." Five or ten minutes later an eagle would glide into sight.

Nicholson's identification system had a particular impact on Sutton. Nicholson rarely mentioned the traditional field marks when describing birds; he seemed to look right past them and to identify each bird by the aesthetic impression it made and the emotion it stirred in him. The words he used to describe the hawks he saw were the same words he used to note the components in a landscape he wanted to paint: "elegance," "compression," "power," "grace." A red-tail was identified not by the belly band the books told you to watch for, but by the "rippling strength" in its flight; a kestrel was identified not by the mustache markings on the side of its face, but by its "delicate buoyancy." And Nicholson's observation of any hawk only began with the identification of the species. He studied each hawk to find its individual characteristics, noting the slightest differences in color or shape. When Sutton asked him to describe a bird he had seen, Nicholson would never say how he knew it was a red-tail or a goshawk, but only how the individual bird he had seen had differed from the other members of its species. "The purple red-tail was soaring over Town Bank again yesterday," he'd tell Sutton, or "I saw a tubular goshawk at the Beanery today."

Sutton found himself imitating his mentor in the field in all ways but one. Nicholson, the artist, was indifferent to record-keeping, believing counting and tabulating interfered with careful perception. He kept no notes on the numbers of hawks seen, no dates of big flights or rare sightings; he didn't even count the owl and hawk nests he found each spring. Sutton, the young scientist, was immediately intrigued with the idea of record-keeping. He memorized the raptor records in Stone's *Bird Studies at Old Cape May* and in Choate's update: early arrival dates, biggest flight dates, even the names of the observers.

By 1975 Sutton was arranging his work schedule at the

Health Department around the hawk flights, working without breaks all spring and summer so he could take his vacation and overtime compensation days in September through November each time a cold front blew through. When he guessed wrong and was caught at work as the winds turned northwest, he spent the day in agony, wondering what he was missing down at the Point. On weekends, no matter what the weather, he monitored the flights, counting the hawks and charting the winds from each spot he and Nicholson visited.

In the summer of 1976 Nicholson identified a new adversary: the New Jersey Audubon Society had announced plans to establish a Cape May Bird Observatory at the Point. The idea was, in Nicholson's view, one more example of outsiders imposing their will on locals. The only local who seemed to have been consulted about the arrangement was Ernie Choate, whom Nicholson did not trust. Worse, the man named to be the Observatory's director was Bill Clark, a bird bander from Virginia who had been running the Cape May hawk-banding operation which Nicholson had bitterly opposed since its inception in 1966. Finally, even more irritating to Nicholson, the first announced project of the Observatory would be a season-long hawk count at the Point.

Although he had never bothered to count the hawks himself, Nicholson knew that neither of the two most recent season-long counts, both conducted by Ernie Choate, had been accurate. Choate was not a particularly dedicated hawkwatcher. He didn't start counting until after he'd had a leisurely breakfast, usually stayed at his watch for only a couple of hours, and generally extrapolated from the first few hours to arrive at an estimate of the total for the whole day. He was frequently unable to get out into the field at all, and had to ask Nicholson or one of the banders for an estimate of the day's flight. His count for the 1965 season had been only 3,951; his count for 1970 had been a more realistic 41,021, but the majority of those birds had come on a single day: October 16, when Choate, using extrapolation, had estimated the kestrel flight at 24,875. Subtract that day's guesstimate from Choate's seasonal total (as most people did, considering it as either a fluke of the winds or a glaring error), and Choate's average for his two counts was barely 10,000 hawks a season.

Nicholson felt sure a more dedicated and perceptive counter would arrive at a much higher total, higher than anyone suspected, and that the political consequences for environmentalists could only be bad. What would it mean to the general public if New Jersey, the most densely populated state in the country, suddenly produced a hawk census that seemed to demonstrate its raptor population was large and healthy? Wouldn't the average person, not understanding the funneling effect of the Point or the logistics of migration, see a high count at Cape May as evidence that environmentalists had been exaggerating their cries of pain?

Hawk Mountain's osprey counts were already being misused in this way. Pesticide interests regularly pointed to the osprey totals at Hawk Mountain, which had been steadily increasing since the 1950s, as proof that DDT was not the villain environmentalists said it was. How could DDT be killing ospreys, they asked, if the best-respected hawk count in the country found more each year? Environmentalists knew the answer, but it was a complicated one and hard to explain in the public forums where conservation battles rage. The national osprey population had been devastated by DDT, and nesting surveys confirmed this fact, but the hardest hit had been the birds nesting on the Atlantic Coast, where thousands of pairs had disappeared. These birds migrated along the ocean and had seldom, perhaps never, been among the birds counted by the watchers at Hawk Mountain. The Hawk Mountain ospreys represented a small minority of the species, an interior population that had apparently escaped the worst of the DDT poisoning and was now probably benefiting by the lack of competition from coastal ospreys on their wintering grounds. Osprey counts at Hawk Mountain had also increased because of the increased attention and sophistication of the watchers at the lookouts there, who grew more intent each year on setting new records. This drive for new records, Nicholson believed, worked against the birds' best interests.

Nicholson foresaw the same kind of thing happening at Cape May Point. How can things be as bad as the conservationists claim, developers and polluters could argue, if ugly, industrialized New Jersey has such huge hawk counts?

Clay Sutton, for his part, had a simpler reason for not liking CMBO's new presence at the Point: he found it insulting that neither he nor Nicholson had been asked to serve as the hawk counter. The fact that the counter who *had* been chosen was a young man from North Jersey who was totally unknown in Cape May was salt in the wound.

The census began on September 1, 1976, and by chance, Sutton was an important witness. He was sent by the Health Department that morning to collect water samples from the Point's beaches, and he started at 9:00 A.M. at the beach off Alexander Avenue, a mile west of the Lighthouse. He parked his truck, waded into the ocean in his bathing suit, scooped up a glassful of water, turned to go, and found himself staring at a peregrine falcon pumping right overhead and out to sea.

Sutton was delighted. September 1 was a new earliest date ever for peregrine falcon at the Point, and better still, his spotting the bird showed who *should* have been selected to do the count. He called Nicholson that evening to tell him about the sighting and share an insider's laugh. "I'll bet the kid on the count missed that one," said Sutton.

Pete Dunne, "the kid on the count," was twenty-four years old, only three years younger than Sutton, and he had not missed the peregrine.

Beginnings
❖

He had spotted it flying west over the beach a couple of minutes after nine, and had duly recorded it. It had been the single highlight in a long and awkward first day, which he had spent standing on a wobbling platform he had built of plywood and two-by-fours, counting hawks while beachgoers strolled past and gawked at him.

The low point of the day came that evening when Dunne returned to the house he shared with the banders, told them about the falcon, and found that none of them believed he knew what he was talking about. "Peregrines are *rare,*" one told him, "and it's too early still. You won't see your first peregrine here until the end of the month."

Dunne was in no position to argue. He had never birded in Cape May before and had seen only a dozen peregrines in his life. Under "Occupation" on his most recent tax return, he'd called himself "a professional ne'er-do-well." Since graduating from college he had worked as a carpet installer, quit that job to enroll at the University of Alaska, dropped out after eight days, and come home again to Whippany, New Jersey, without a job or a clear ambition. Working again part-time as a carpet installer, he had found himself increasingly addicted to hawks and through September and October of 1975 drove the 220-mile round-trip from Whippany to Hawk Mountain two or three days a week.

In the spring of 1976 Dunne had talked New Jersey Audubon into hiring him to conduct a two-month census of the spring hawk migration from Raccoon Ridge in the New Jersey Kittatinnies. The pay was $340 for the season. He lived in a tent and a down sleeping bag at the top of the mountain from March to May, skipped lunch daily, and served himself the same meal out of the same pot every night: Kraft macaroni and cheese, mixed with water. Hawks were few, and one day the wind meter topped out at seventy miles per hour, but when New Jersey Audubon told him about their plans for the new Cape May Bird Observatory and asked if he'd like to be the

[47]

official hawk counter there, Dunne said yes. The pay would be $500 for the season, a raise he calculated would enable him to upgrade his nightly meal to canned ravioli.

He soon learned that the proposed observatory had generated opposition throughout the state. Dunne's most constant companion on Raccoon Ridge, a veteran hawkwatcher named Floyd Wolfarth, whose guidance had meant as much to Dunne as Nicholson's had to Sutton, erupted in anger when he heard Dunne had agreed to participate. "I *forbid* it!" Wolfarth bellowed. Like Nicholson and other veteran raptor watchers, Wolfarth had long opposed the banding operation's presence in Cape May. Banding he associated with falconry, and falconry he considered despicable. Falconry in many birders' minds was an even worse sin than hawk shooting. Unlike gunners, who were ignorant people for the most part, falconers were educated naturalists who professed to love hawks, yet their sport involved removing birds from the wild and imprisoning them— merely for personal pleasure. In Wolfarth's view, banding was this same, poisonous wine in a new bottle. Some birders suspected that the entire Cape May banding operation was a ruse, an elaborate sham designed to enable falconers to trap birds to take back to their cages at home, under the cover of a scientific enterprise—which New Jersey Audubon was now legitimating by establishing its new observatory.

Dunne arrived in Cape May that fall to find that the observatory was still more concept than reality—he had no office and lived in an unheated front room in a house the banders rented—and that the banders were as wary of their new connection with the birding community as the birders were of them. Several of the banders *were* ex-falconers, and they knew this made them unredeemable sinners in many birders' minds. The banding operation had been an independent, self-funded enterprise for ten years, and working from dawn to dusk in blinds that were inaccessible to birders and the general public, the banders had become a very close-knit group. Some didn't trust Dunne at first—for all the banders knew, he could have been a spy sent into their midst to catch them in some error—and the rest were confused by him. Why didn't he want to become a bander himself? He obviously loved hawks. Why did he only want to count them—when he could be trapping them and holding them in his hands?

One morning in September, a couple of weeks after his peregrine sighting, Dunne learned of still another kind of opposition to his presence at the Point. A tall, balding man wearing a huge pair of binoculars on a rawhide string walked up to him at his watch. "So here he is," the visitor boomed, "our mail-order hawk counter, that naive young man so duped by his own simplicity he doesn't realize his data will be used to justify the continued decimation of raptor populations everywhere." Dunne could think of nothing to say. "Well, I've got to be going now," Al Nicholson said. "Sorry if I spoiled your day." And he turned on his heel and walked off.

For a month, Dunne was a man without a camp. But one day Choate mentioned to Nicholson that Dunne had claimed he'd seen a peregrine falcon on September 1, Nicholson reported this to Sutton a few days later, and Sutton soon told Choate that "the kid" had been correct: a peregrine falcon *had* flown by that morning. One afternoon Sutton stopped by Dunne's platform to say hello, and after an hour or two of hawkwatching, they drove to the C-View Inn to have a beer. They have been close friends ever since.

"Everything started to change," Dunne will tell you today, "when Clay Sutton stepped forward and confirmed that Day One peregrine. He was the first local to accept me."

Sutton has a more modest recollection. "It was October by the time I finally said anything about that bird. I didn't like the idea of some hotshot from North Jersey coming down into my backyard to tell us what hawks we had. Even after Al and I got to know and like Pete, we spent most of the next two falls competing against him. We'd wander over to the Point, get him talking about what he'd seen, then walk off and laugh about it: 'He thinks he's seeing hawks. *We* know where the hawks are.' It was ridiculous. Al and I knew the winds and the flight paths, and we could move around, playing the weather. Pete was stuck at that one spot at the Point—and still he was seeing a lot more birds than anyone had ever imagined he would. Nowadays he's become so well known as a personality that people overlook how good a birder he is. What impressed me most those first few years was how fast he picked things up. One reason the count went from forty-eight thousand the first year to eighty-one thousand the next was because Pete improved so much as a hawkwatcher. That first year he was depending on the traditional field marks. By the second year he had adopted Al's methods, what we call the 'holistic' system now, and he was really pushing against the envelope of identification, calling birds he'd had to let go the first year."

Dunne borrowed a lifeguard chair from the Point Beach Patrol for the 1977 and 1978 counts, and it became the primary gathering spot for birders at the Point. Visitors dropped by to hear what rarities were around, to report their own sightings, and to try to test their identification skills against Dunne's. The conversation and competition were so good that many stood by his side all day long. Stories, jokes, and quips rolled off Dunne's tongue as steadily as the hawks flew by. He had renamed most of Cape May's avifauna. A great black-backed gull was the "Imperial Dump Buzzard," tree swallows were "Merlinettes," the black-crowned night heron was the "Squawk." A bird so far out even Dunne couldn't call it was "beyond the limit of conjecture." When anyone dared challenge Dunne's identifications, he always had a comeback. "You might be right," he'd say, "but *I'm* official." He invented an organization, the Brotherhood of Professional Hawkwatchers, declared that anyone interested in joining had to prove himself to the membership first, and designed an arm-patch: a hand holding binoculars manacled to the wrist and aimed at a peregrine flying by the Cape May Lighthouse; the Latin script beneath the falcon read *veni, vidi, computavi,* "I came, I saw, I counted." Earning that arm-patch soon became as important a quest among hawkwatchers throughout the East as spotting the next gyrfalcon.

By the third year of the count Dunne had become director of the Observatory and was proving his abilities on other fronts—as writer, photographer, and speaker. He was on the road, giving slide shows throughout the state, two or three nights a week. The Observatory's newsletter, *Peregrine Observer,* grew thicker each issue. What had been six pages of mimeographed announcements became a twenty-eight-page journal, featuring a smorgasbord of articles—lyrical evocations of the Cape May scene, angry essays on the abuse of insecticides, tongue-in-cheek fictions and fables about birder's dreams and nightmares—virtually all of them written by Dunne. Soon he was writing essays for *Birdwatcher's Digest,* Cornell's *Living Bird Quarterly, American Birds,* and finally for a biweekly column in the New Jersey section of *The New York Times.* The best of his *Peregrine Observer* pieces were eventually published by Rutgers University Press in a collection entitled *Tales of a Low-Rent Birder.*

"People ask me now," Sutton says, "how could it all happen so fast? In nineteen seventy-five, there were maybe a dozen people watching hawks at the Point. Three years later there were hundreds, and by the mid-eighties, we had thousands. The counts published in *American Birds* and the Hawk Migration Association newsletter had something to do with it; that's what brought the raptor nuts. But most people who came weren't fanatical hawkwatchers, and they weren't coming only to see the hawks. They came to see Pete Dunne in action. They read his stuff, and they wanted to meet him, talk to him, listen to his stories. As far as I'm concerned, Pete Dunne made Cape May Point hawkwatching what it is today. We had always had the birds. What we didn't have before nineteen seventy-six was Pete Dunne."

Not all who met Dunne liked what they saw. He was tall, handsome, and smart—and knew it. And he missed few opportunities for self-promotion. One year he designed a full-page ad in *Peregrine Observer* soliciting sponsors for a spring birdathon. Half the page was given to a photograph of Dunne dressed like James Dean in sunglasses and T-shirt. Above it ran the largest print his *Peregrine Observer* ever used: THIS IS A BIRDING MACHINE! The next year he had himself photographed in the pose of a Unicef Starving Child, sitting next to a trash barrel holding up an empty bowl. "You can help Peter," read the copy, "or you can turn the page." When Dunne and his two birdathon teammates saw 185 species in one day, breaking the twenty-year-old state record of 176, some observers were more irritated than impressed. The following year Dunne used the back page of *Peregrine Observer* to feature a photograph of himself and his two teammates raising a tripod in imitation of the Iwo Jima Marines. The Guerrilla Birding Team, as they now called themselves, would not only break the state record again, Dunne announced, but would find 200 species in a single day. Only Texas and California had single-day records that high. "The experts say 200 species is impossible," Dunne wrote, "the Guerrilla Birding Team thinks they're wrong." That May, when Dunne and his partners found 194 species, breaking their own record by 9, rebreaking the old record by 18, but missing their goal by 6, his critics saw the event as a defeat for him and a victory for modesty and restraint.

Modesty and restraint were not Dunne's forte. One year a local critic of the banding project wrote to Hawk Mountain Sanctuary claiming that she had come upon a hawk entangled in an untended net near the Meadows and that hawks trapped at the Point were regularly mishandled. Maurice Broun wrote to Bill Clark asking if there were any truth in the letter. Clark fired back a response, inviting Broun to come to the Point, tour the banding stations, and see for himself how the hawks were treated.

Broun arrived a few weeks later, undercover. Dunne saw him in the parking lot, and introduced himself. When Broun refused to give his own name, Dunne guessed what was going on, and tried to lure him out. "Where are you from?" asked Dunne. "Pennsylvania," said Broun. "Have you ever been to Hawk Mountain?" Dunne asked. "I've been there," said Broun. "You know," said Dunne, "the one person I'd most like to meet is Maurice Broun. He's my idol. The best hawkwatcher in the world, and the Keeper of the Flame at Hawk Mountain. Someday I'm going to drive up to the mountain just to shake his hand and tell him how much he's inspired me." Broun wouldn't take the bait, however, and eventually walked away.

That night Dunne stopped by a birder's house and found Broun sitting in the living room. The reconnaissance was over, so Broun could come clean. He stood up. "You know, young man, today you said that one person you'd like to meet is Maurice Broun. Well, I want—"

Dunne cut him off with a dismissive wave. "Oh, sit down, Maurice. I knew who you were."

But despite the brashness—or maybe because of it—Dunne was a masterful creator and promoter of ideas. To raise funds for the banding project, he created Project Wind Seine, which sold certificates of sponsorship of individually banded birds and brought in thousands of dollars for the Observatory. The project has proven so successful that it is still today the sole source of funding for the banding project and has been imitated by other bird observatories across the country. To generate interest among younger birders, Dunne organized a High School Hawkwatch Program, where students ran their own hawk counts at their schools, and he wrote a manual, *Hawk Watch: A Guide for Beginners,* to help get them started. To answer the questions about duplicate counting from the platform, Dunne designed and ran an Expanded Hawk Watch for two seasons, which involved recruiting half a dozen different hawk counters, deploying them at five different observation stations around the Point, and coordinating their counts.

One afternoon, sipping beers with some friends at the C-View Inn, Dunne got to wondering out loud whether his birdathon team needed some human competition to push them past the 200-species mark. It seemed to Dunne that in neither of their two record-setting days had they been pressing as hard as they could have been. What they needed, Dunne said, was the threat of other teams out in the field that same day trying to outdo them. These other teams would have to be good, of course. How about issuing a national challenge in *American Birds* and the other birding magazines daring other teams to come test their skills and luck in a head-to-head competition? The rules would be simple: twenty-four hours of birding inside New Jersey's borders on a designated day in May. "We could call it," Dunne announced, "The World Series of Birding!"

Almost everyone who listened to Dunne's spiel over the next few weeks had the same question, "How could a New Jersey birdathon declare itself The *World Series* of Birding?" Then Dunne had another inspiration. He telephoned Roger Tory Peterson, the world's most famous birder, in his home in Connecticut, introduced himself, described his idea, and explained that the intent of the competition would be to raise funds for conservation causes, so the hype in the name was necessary. Would Peterson be willing to support the idea by endorsing the concept?

"Whose team can I be on?" Peterson said. "Can I be on yours?"

That, Dunne wrote later in *Peregrine Observer,* was "like

having the Pope ask you whether he can go to church with you on Sunday."

On May 19, 1984, Dunne chauffeured Peterson and the rest of his team to victory in the first World Series of Birding and to a record 201 species for the day. The climax came in the late afternoon on Bayshore Road in Cape May when they spotted a fork-tailed flycatcher feeding in a farm field—a species so rare that it was a lifer for the Grand Master himself, Peterson's 697th North American bird. The event made the national news, and the World Series of Birding has grown larger every year since, drawing teams from California, Canada, and Great Britain, and raising nearly a million dollars in pledges for conservation causes.

Those hoping to see Dunne's balloon burst were disappointed. He kept rising higher and higher. When Simon and Schuster purchased the rights to republish *Tales of a Low-Rent Birder* in a paperback edition, they added a subhead in sharp red print on the front cover: "19 Flights of Fancy by America's Second-Best-Known Birdwatcher."

Dunne's last project before he left Cape May in 1987 was to write a book with Sutton and David Sibley, one of Dunne's World Series teammates: *Hawks in Flight: The Flight Identification of North American Migrant Raptors.* Dunne and Sutton had been discussing the idea almost as long as they had known each other—to put in print the identification techniques they had learned from Nicholson and Dunne's first mentor, Floyd Wolfarth, and then developed on their own over more than a dozen seasons.

Sutton played Dunne's anchor man through his Cape May years, charting the hawks each year, keeping the records of sightings and flights and dates, pushing Dunne on conservation issues, playing the hard-eyed realist to Dunne's garrulous Quixote. He had trained himself as a photographer, and his photographs regularly illustrated

Dunne's articles and eventually their book. He was also CMBO's friendliest critic, keeping Dunne in touch with the local community and with the scientific issues beyond the public relations. When they realized they knew too little about western raptors to write convincingly about them in their book, it was Sutton who drove west to photograph them and fill notebooks with descriptions.

Dunne is gone from the Cape May scene now. He has moved back upstate to work in New Jersey Audubon's office in Bernardsville, as director of natural history information and editor of the society's magazine, and has moved on to wider adventures, leading trips to Australia, Alaska, Israel, and elsewhere. Frank Nicoletti and each of the other hawk counters who have followed Dunne have also moved on after two or three seasons. Al Nicholson is seldom seen around the Point anymore, having retreated more deeply into solitude. Ernie Choate has passed away.

Sutton remains. He has witnessed and charted every season's hawk flight since 1974, watched more hawks fly through Cape May than anyone else, and has become the hub of the wheel of the hawkwatching culture at the Point.

One evening in August Sutton stands on the back porch of his house in Goshen, ten miles north of the Point, where he and his wife, Pat, live with their two English setters. Swallows, egrets, gulls, and ibis are winging by overhead, heading east, away from the sunset toward roosting areas inland. Sutton bends his neck to watch each flock until it flies out of sight over the house. His upright, shoulders-back stance and thickset torso suggest an owl's. His eyes are bright and unceasingly attentive, his face is round, and he seems to listen with his whole body, turning to face and give full attention to anyone talking to him. He laughs quickly and frequently, most often at himself.

In the Suttons' garden half a dozen hummingbirds chase each

other, chitter-chattering over the bee balm, salvia, and coral bells planted for their benefit. At the garden's center are three sunken bathtubs where, over the years, Pat has counted as many as fifty frogs of six different species, though the nearest stream is three hundred yards away. A moth-rearing cage, two feet by three, is open and empty next to the porch; above it, cecropia moth caterpillars, each as long and thick as a linebacker's thumb, feed in a cherry tree.

Clay and Pat met in 1976 at a party of his cousin's. Clay was standing in the kitchen when a young woman entered wearing a leopard skin coat. "I was deep in my Al Nicholson phase in those days, thinking confrontation was the only way to reach people, and I blasted her. She told me the coat was sixty years old and her grandmother's hand-me-down. I said that made no difference at all." A month later he invited her to Florida on a birding expedition, and she agreed to go, though she had never identified a bird in her life. A year later they were married, and today Pat Sutton is one of the most influential conservationists in South Jersey. She is a CMBO's teacher naturalist, the editor of *Peregrine Observer,* a skilled cartographer and photographer, the author of numerous articles on owls, hawks, butterflies, and native plants, and a relentless environmentalist. Her years of work to preserve the wild areas of Cape May County, especially the South Cape May Meadows, won her the 1986 President's Stewardship Award from The Nature Conservancy as the outstanding conservator in the nation. Over the last few years she has been the principal grass-roots organizer of the effort to establish a Cape May National Wildlife Refuge, an attempt to preserve and protect the Delaware Bay. "Yeah," says Clay, "but I still haven't talked her into throwing away that coat."

Their house was built in 1830. It is small and as crowded as a nature museum. At the moment, an elk antler, five feet long, dominates the living room, resting among stacks of journals and bird books on the coffee table. Clay found it this summer while he and Pat were tracking goshawks and Swainson's hawks in Canada, and he has yet to decide where he can clear a space for it. Duck and shorebird decoys stand on each windowsill and on all tables in the room: a black-bellied plover carved by Clay's grandfather; a trio of low-slung, impressionistic mergansers by local carver Tony Hillman; a wigeon woven from straw in Italy; and black ducks in various styles, from 1920s working decoys to contemporary decorative birds. Purple catfish netted in a nearby stream swim in an aquarium at the entrance to the den. The floor-to-ceiling bookshelves inside hold more than a thousand volumes—two walls of bird books and another of fishing books. Clay's fishing rods are suspended from the ceiling; slide carousels are stacked at the doorway.

The stairway in the living room holds fish prints, photographs of raptors—red-tail, peregrine, golden eagle, snowy owl—and a four- by six-foot aerial photograph of Cape May Point. Six or seven canvases on the living room's center wall are by Al Nicholson, all of them dark and vibrant landscapes painted in South Jersey. In one an owl perches on a tree stump at dusk; in another a hawk looks over a rusting piece of farm equipment. The few human figures are shadows.

Nicholson and Sutton had a falling out a few years ago. The source of their conflict was Bear Swamp, the first piece of wilderness they had walked together. A coalition of conservation groups was urging the state to purchase the forest around New Jersey's last active bald eagle nest, which was threatened by woodcutting and sand-mining. Nicholson, representing CAPE, thought Sutton, representing New Jersey Audubon, was too compromising. Sutton thought Nicholson was driving their potential allies away with his heat. Bear Swamp was saved—it is now a protected forest, the last fifteen hundred acres of coastal-plain climax forest left in New Jersey: Sutton's and Nicholson's friendship was broken. Today, when they meet, they exchange hellos and an occasional piece of raptor news, but

Sutton no longer spends the day at Nicholson's side watching the sky while Nicholson paints, and Nicholson no longer arrives at Sutton's house with a pile of books of Civil War photographs to show him.

"I miss him, and I miss his perspective," says Sutton. "He taught me hawkwatching, and he had a big effect on Pete too—showing us both how much more you could see by developing your intuitive sense of birds. I can't tell you how much of Al Nicholson is in our book. If Al had been a writer, he could have written that book twenty years ago.

"And the older I get the more habitat destruction I witness, so the more I come to understand Al's anger about what's going on. Sometimes when I really get feeling bitter, seeing some new mall or condo development going up, I think, 'Al's been watching this destruction get worse and worse every year since the 1930s.'"

Sutton's job with the Cape May County Health Department continued for a dozen years, until he had thirty-three people reporting to him, in sixteen different programs. "It was the typical middle-management position: all the responsibility, none of the authority." His stomach was tied in knots at night. One July afternoon in 1986 he felt so fidgety and sick during a meeting that he asked a coworker to drive him home. His arms hurt, his chest hurt. He thought it was a bad case of the flu, went to bed, and woke up the next morning, feeling worse. His wife took one look at him and drove him to the doctor; the doctor took one look at him and sent him to the hospital. Sutton had had a heart attack, at age thirty-six.

"As everyone says who ever had something like that happen, it put things in perspective. I was in the hospital two weeks, took a four-week vacation, and then resigned. One night after hours I drove to the office and cleared out my desk. I didn't want to see my boss. I was afraid I'd punch him in the nose. I blamed the whole thing on him."

He works now for an environmental consulting firm, Herpetological Associates, a job that allows him to spend time outdoors several days a week and to work at night when the flight at the Point keeps him there.

"The CMBO count starts officially on the fifteenth of August, but I start earlier. We usually get an eagle or two in July, though I don't count them as fall migrants because we don't know where they're going. Then, in early August, the first few broad-wings and kestrels come through. I always think of the early season and the late season as the best times for something off the wall showing up. Some years we've had Mississippi kite, and I'm always thinking about the possibility of a swallow-tailed kite in fall. We get one or two swallow-tails every spring now, but the only fall record is September 1, 1946, by J. d'Arcy Northwood. It was his first day as Audubon coordinator here, and he saw a swallow-tailed kite. More than forty years later, we're still waiting for the next one."

Of the twenty-four species of hawks on Cape May's official all-time list, Sutton has seen twenty-three, more than any other observer. He has missed only the European hobby, seen by Frank Nicoletti on its single recorded appearance on September 28, 1987. He also counts as a miss a black-shouldered kite a visiting birder reported in October 1972. "That record has never been accepted, but I have his notes at home and a sketch he did in the field. The drawing shows the commas in the wings. I think it was a legitimate sighting, and it's a species we can hope to see here again.

"I have a couple of might-have-beens, too. Early one morning during Frank's second season, he and I were on the platform when the banders radioed and said they had a golden eagle way up high. We looked straight up and there was the bird—already up in the stratosphere and it wasn't even nine A.M. You don't usually see eagles up that high that early, and this one didn't look like a golden *or* a bald. The wings were *dark;* the only white we could see was in the undertail coverts. We watched it for maybe a minute while it circled

around and then crossed. Later, Bill Clark, who was the one who'd
called it a golden, said it hadn't looked right to him, either. So I asked
him, 'How about spotted eagle?' He's seen them in Israel on migration. After about two minutes of silence, he said, 'You know, I never
thought of that.' It's a bird I think we could expect in the U.S. because
eagles will cross water, and Frank says now it *was* a spotted eagle,
but neither of us has written it up, so I guess we both have doubts.

"The other bird I wish I'd had another look at was a little
falcon that came down the railroad tracks one day in September five
or six years ago. Pat and Pete were with me, but sitting down. I
shouted, they jumped up, and the bird went by no more than seventy
feet from us. It was black and white and *tiny,* and I think it was very
possibly a red-footed falcon, which means our sighting would have
been a North American record. But it was gone almost as soon as I
said anything, and Pete no longer admits he saw that bird. 'Credibility
is a delicate thing,' he says.

"The species that everyone has been predicting will show up
next is prairie falcon. It's the only temperate-zone North American
hawk that *hasn't* been reported in Cape May yet, but it's a short-
distance migrant, and I keep wondering if we might actually have
a better chance for a black kite. Black kite is the most numerous
raptor in the world, nests in Europe and Asia, and migrates over
water, long distance, to Africa. There was a report of one in Oregon
a couple of years ago."

Sutton is drinking a beer. He takes a sip, then studies the sky
for a minute in silence before speaking again. The ibis and gulls are
still winging by, flying away from the reddening sky. "Whenever I
get talking to other people about the flight, I find myself talking
about the rarities and the big flights. But whenever I *think* about the
flight, it's not about those things. Rarities and big flights are the 'Oh,
Wow!' part of the phenomenon. Those are the components that get

you interested at first, and they're the easiest components to talk about. But you won't find many veteran hawkwatchers who go out to their local lookout because they're hoping to find a black kite or a gyrfalcon. It's *curiosity* that keeps you watching year after year. And the more years that go by, the more you know your particular watch, and so the more curious you are. When is the first harrier going to come through this year? How many ospreys will we get? Will peregrine numbers keep climbing? How are the sharpies and kestrels going to do this year? I've been more and more worried about those two lately, the sharpie especially. Our sharp-shin count has been dropping steadily since the early eighties. Will this be the year that breaks the trend?

"And curiosity isn't all there is to it, either. There's also that feeling of awe I think every serious hawkwatcher feels, even on a bad day. Watch one bird flying south and you can feel it. *How* does that hawk know where she's going? And how long have hawks been flying on that same route? Al used to call it 'the pageant and the mystery,' and that's what it is. The pageant and the mystery are really what keep you going. Pete had a line about it I like to quote. He was writing about a peregrine flying south its first fall. How was she feeling and where was she going? She was 'restless to return to a place that she had never been but would recognize once she got there.' "

Sutton pauses for another look at the sky. "I can't stay away from the flight now. Every season you get under your belt makes the next season more meaningful. They blur together in your memory after a while, but that doesn't matter. The longer you study migration the longer you want to. August comes around each year, and I can't imagine being anywhere else but at Cape May, watching hawks come by."

Chapter Three

❖

MORNING, NOON, AND NIGHT

THE FIRST HAWK OF THE NEW MONTH, A KESTREL, PUMPS PAST THE
platform, heading out over the Bay, at a few minutes before 9:00
A.M. on September 1. "I'll be damned," says Jeff Bouton. "Hawks
really do migrate through Cape May."

    The season has not begun well for Bouton. Elsewhere around
the Point the birding has been excellent: in the Meadows, a half-mile
east of the platform, a dickcissel, a clay-colored sparrow, and four
sedge wrens have been found; at Higbee's Beach, a mile and a half
northwest of the platform, Richard Crossley has counted more than
eight thousand warblers in three weeks, including twenty-four hundred in a single morning; last week at the Second Avenue jetty a
birder saw two seabirds he identified as black-capped petrels, a deep-
water species that has rarely been reported north of the Carolinas and
never from the New Jersey shore. Stuck at his post on the platform,
Bouton has seen none of these birds. His best find so far has been an
insect—a snout butterfly which rested briefly on the ramp's railing
yesterday afternoon.

August's total was 268 hawks. An average count for the month would have been closer to 500, and a good count would have broken 1,000. The only decent flights so far came on Bouton's first two days on the platform—forty-eight hawks on August 17 and eighty on August 18. Since then, most daily counts have been in the single digits, and Bouton has been shut out completely on five different days. It's too early to make projections for the season, but like a teacher whose class has flunked their first quiz, Bouton is glum. "I might have missed a hawk or two in the last couple of days, but after you've been staring at the sky for eight hours without seeing a single migrant, your attention span isn't too great." Bouton scans

" . . . Straight up so high they're invisible."

in S-shaped sweeps, first low from left to right—wooden tower, flat-topped oak, round-topped oak, round-topped pine, merlin sticks, radio towers, water tower, dunes, bunker—then he raises his binoculars half a field and moves from right to left—bunker, dunes, water tower, radio towers, merlin sticks, round-topped pine, round-topped oak, flat-topped oak, wooden tower. By the third sweep he is above the tops of the trees, and he continues upward until he is looking almost directly straight up. Then he comes down to the horizon and begins again.

An osprey comes in off the ocean, carrying a fish. "It's only a local," says Bouton. "A couple of pairs nested in the Meadows."

"Can you I.D. the fish, Jeff?" asks Vince Elia, one of the few birders who has bothered to visit the platform in recent days.

Bouton lifts his binoculars. "That's a menhaden," says Bouton. "It's dorsally compressed."

"Right," says Elia, "and they bleed more profusely."

"Remember the needle-nosed gar last year?"

"Yeah, that was neat."

"At Sandy Hook," says Bouton, "I saw them carrying winter flounder."

"Yeah, I've seen that," says Elia. "I've seen them carrying goldfish."

"A guy I know from Rochester told me he once saw one carrying a chipmunk."

"Oh, yeah?" says Elia, his eyebrows lifting. "Was he drunk at the time?"

Twenty minutes pass without another hawk, and Elia leaves the platform to try his luck tracking down the Connecticut warbler reported at Higbee's Beach. Bouton watches him walk down the ramp. "If you see it," he calls, "come back and tell me." Elia waves and drives off.

Bouton's only regular companion in the last week has been a gray tree frog, which has climbed up one of the posts and taken up residence in the little crevice above a DO NOT PERCH ON TOP RAILING sign. As the sun reaches into its hiding place, Bouton plucks the frog out and carries it down the platform to the next sign, which still has shade. The little frog has coffee-brown mapping on its back and long fingers knobbed like ET's. "There you go, buddy," Bouton says as he inserts it rear-end-first in the tiny space. "Don't want you to get desiccated."

Bouton sits down and, head bowed, scrapes at the deck absently with one sneaker. His binoculars hang limp.

It is the afterimage of his predecessor Frank Nicoletti that keeps Bouton at his post. Pete Dunne and the other counters who served at the Point before Nicoletti simply walked away from the platform on slow days, especially in August and early September when the flights are always spotty. The rationale was simple: the five or ten hawks you might count on a slow day's watch in the summer sun were meaningless in a season-long total of fifty or sixty thousand hawks. More important was saving your eyes and energy for the good days of late September and early October when you might count five or ten hawks in a couple of minutes. To Nicoletti, "The Iron Man," however, a season on the platform was primarily a test of stamina and brute persistence—more like a pole-sitting contest or a hunger strike than a bird census. He knew the total of hours spent watching was a better measure of a counter's dedication than any spectacular flights he might happen to witness. The flights are erratic, and record counts come only when weather, winds, and seasonal timing happen to match up in lucky combinations; the hours come by one after another—at the same steady pace no matter what the circumstances.

Of all Nicoletti's records, his string of three consecutive thousand-hour seasons is the one least likely ever to be matched. To total a thousand hours in a season Bouton would have to average eight hours on the platform each day, seven days a week, for eighteen weeks straight. He knows already he cannot do it. None of Cape May's hawkwatchers has ever matched Nicoletti's monomaniacal passion for raptors. But Bouton knows also that his effort will be measured against his predecessor's by all the Point's insiders, and he wants to come as close as he can. So, even on a day like today, when the hawk total will be meaningless, he stays at his post on the platform.

He lies flat on his back on the bench and studies the sky directly overhead. His binoculars move in slow sweeps. "There's a plane," he says after a long search, "what a relief to see something." He rests his binoculars on his chest, closes his eyes, and yanks his cap down over his face. Five minutes later, he pushes the cap back, reaches for his binoculars, and returns to scanning the burning blue sky.

"That's where the hawks are on days like this," he says, "straight up so high they're invisible."

"He's right," says Paul Kerlinger, the director of the Cape May Bird Observatory. "On a day like this, you need radar to count hawks." Kerlinger rocks in his chair, tapping his pen on his desk in his office on East Lake Drive, half a mile from the platform. Pete Dunne's successor is a tall, thin man of forty, who shares CMBO's crowded, unheated office with Pat Sutton, two other staff members, half a dozen volunteers, two dozen hawk banders, and countless ever-present visitors. His binoculars stand upright like a paperweight on a stack of ornithological journals on the windowsill behind him. Gulls and geese float on the lake across the street. Mail, phone messages, notes, and photocopied articles cover Kerlinger's desk. The four tallest piles of papers are on the floor, two on each side of his chair.

A photocopier occupies one corner of the room; a tilting, overloaded bookshelf occupies another. Pat Sutton's neatly organized desk contrasts with Kerlinger's and squeezes tight against it; a computer sits on another desk in the room's center. The office's main door opens to a small vestibule where the Observatory sells field guides, T-shirts, patches, and visors. The other door opens to a narrow kitchen. At a makeshift desk in the back, beyond a counter top piled high with mugs and bowls, is a second computer.

On Kerlinger's lap is the final draft of his book, *Flight Strategies of Migrating Hawks,* about to be published by the University of Chicago Press. The manuscript is as thick as an auto parts catalog and has just come back to him from the press's copy editor spotted with red marks. "She can correct my style," he says, flipping the

*Cape May Bird
Observatory entrance*

pages. "But I'm not changing the graphics." The book is filled with graphs and charts: a graph comparing the maximum gliding distances of monarch butterfly, broad-winged hawk, albatross, and motor glider; a chart comparing the flocking tendencies of 133 species of hawks worldwide; another summarizing the glide direction of raptors tracked by radar over two autumn migrations and three spring migrations (1978–80) in upstate New York; a map of banding recoveries of gray-faced buzzard eagles migrating from Japan to the Philippines; a graph plotting the costs of flapping flight and of gliding flight against standard metabolic rate; another plotting altitudes of broad-winged hawks against minutes after sunrise on four mornings during spring migration at Santa Ana National Wildlife Refuge in Texas; a map of prevailing winds in North America in spring and fall. The text is sprinkled with equations. $Q = N/m^2$, for instance, indicates that wing loading can be calculated by dividing the unit of force in Newtons by the wing area in meters squared. Another equation indicates how minimum air speed of a flying bird can be determined by weight, drag, and wing area:

$$V \min = \left(\tfrac{2W}{pC}[Lmax]^s\right)^{\tfrac{1}{2}}$$

Kerlinger is a mathematical man. "If you can't measure it," he likes to say, "it ain't science."

Hawks look like better travelers than they are. Their big wings, long tails, and rounded, muscular chests seem the epitome of avian aerodynamics. Actually, hawks are limited flyers. The expansive wings of eagles and buteo hawks are most useful in slow, circling flight, as the bird searches for prey on the ground. But the larger the wing the harder it is to flap, and so eagles and buteos are poorly adapted for long-distance flight. The rectangular wings and rudder-like tails of the three accipiters (sharp-shinned, Cooper's, and goshawk) are at their best in short, zigzagging pursuits through thick woods. An accipiter pursuing a warbler at full throttle is a blur

through the branches. Out of the woods, accipiters fly like aerody-
namic cripples. Watch the labored, herky-jerky pumping of a sharp-
shin as it crosses a treeless field on a windless day, and you can't help
wondering how it keeps going.

Hawks are designed for predation, not migration. A sharp-
shin may be able to chase down a warbler in a fifty-yard sprint
through bramble and brush, but release both birds from a ship fifty
miles off the coast and it will be the warbler that makes it back to
shore. Almost any small bird can outfly a hawk over the long haul.
Warblers, flycatchers, orioles, swallows, and thrushes regularly fly for
twenty and thirty hours straight, far out over the ocean. Of North
American hawks, only the three falcons have body shapes made for
extended, powered flight—and they and the fish-eating osprey are the
only hawks regularly found at sea. All others avoid long overwater
crossings and fly only by day, when winds and thermals can give them
the extra boost they need.

"Flight is much costlier for heavier birds," Kerlinger ob-
serves. "Weight is a power function, and the numerator in equations
about energy and flight. The heavier the bird, the higher the metabolic cost of flying." On migration, hawks are like football players
forced into a marathon run.

In fact, most hawks do not migrate. According to Kerlinger's
tally, 152 of the 285 Falconiformes in the world are nonmigratory,
living year-round in the same territory. "That's the majority, fifty-
three percent, and only a small minority, six percent—eighteen of
two hundred eighty-five—are complete migrants. The rest are either
irruptive species or partial migrants." Irruptive species move erratically rather than seasonally—usually because prey has become un-
available in their regular territory. Partial migrants are those species
with some populations that do not move: southern populations remain resident year-round, for example, while northern populations

come south; or adults remain on the home grounds while juveniles go south. "Most of the hawks we see at Cape May are partial migrants—red-tail, red-shoulder, Cooper's, sharpie, harrier, the two vultures, the two eagles, and the three falcons. The only complete migrants we see in numbers are osprey and broad-winged hawk."

Kerlinger, who has ridden in sailplanes and motor gliders to study raptor flight at firsthand, is wary of the common comparisons between hawks and planes. "People like to say hawks are better flyers than planes. The aerodynamics aren't that simple. It's true planes can't change their flight profiles by pulling in their wings or spreading their tails, so they can't ride in the tight circles or the weak thermals hawks can ride in. And planes can't put down in a tree or on the side of a cliff when they run out of fuel. But hawks fly much more slowly than planes, even gliders, and their glide ratios are much less. A sailplane has a glide ratio of about forty to one. That means it glides forward forty meters for every meter it loses in altitude. Even an albatross can't match that; albatross glide ratios are about twenty to one. Condors have about a fifteen-to-one ratio. An osprey's glide ratio is about twelve to one."

A gliding bird is always falling relative to the air surrounding it, and long glides are possible only when the air under the bird is rising faster than the bird is falling. In still air all birds must pump their wings to stay aloft.

Because flapping their wings is metabolically so expensive, hawks need winds and thermals to help them cover long distances. This is the reason hawks migrate by day, unlike most other groups of birds, which migrate primarily at night. The air is generally more turbulent during the day; evening and night air is usually too calm for hawks. "It's like the difference between flying a prop plane and flying a glider," says Kerlinger. "Birds that rely on flapping their wings are like prop planes. They want smooth air, so they fly at night when the laminar flow is steadier. In a prop plane you want to avoid turbulence. In a glider, you *want* turbulence. You can't fly very far unless you can find it. Hawks are like glider planes, and that's why you see gliders and sailplanes at Raccoon Ridge and Hawk Mountain: they're there for the same reasons the hawks are. They need that ridge wind lift."

Along the ridges, hawks ride in the winds that are deflected off the hills. Away from the ridges, migrating hawks depend on thermals to carry them along.

Northwest winds generate the largest flights at Cape May and other coastal hawkwatch sites in the eastern and southern United States, but not, as generally assumed, by direct action. The winds do not blow the hawks southward. Northwest winds cool the air and clear the sky and so create conditions that lead to thermal formation. Cold fronts don't push hawks ahead of them; they pull them behind. The ideal traveling weather for hawks comes on the day *after* the passage of a cold front, when the air temperature has dropped and the sky has been cleared of cloud cover. On bright, cold days, the sun warms the ground more quickly than it does the air, the ground heats the layer of air closest to it, and the warmed air rises in invisible, swirling columns.

The hawk flaps upward into the thermal, spreads its wings to ride as high as possible, and then as the thermal dissipates, the bird glides forward along its migration track. The sequence is repeated— slow upward swirl, fast downward glide; another upward swirl, another downward glide—for hours. The daily migration flight of a hawk looks like a child's drawing of bed springs.

The weather that leads the hawks to the Point is not the weather that is best for crossing Delaware Bay, however. "You're a sharp-shin," says Kerlinger, sketching a map on the back of an envelope with his felt-tip pen, "and the wind is out of the northwest.

You're at an altitude of two thousand feet, coming from somewhere inland in New Jersey, riding a thermal and gliding toward the Point. You see the Bay in front of you, and off in the distance, fourteen miles to the southwest, you can see Delaware. That's where you want to go, but it's a long flight across water, and you can feel the winds pushing you the wrong way—left—out over the ocean." Three arrows squeak across the envelope, pointing toward Africa. "What are you going to do? You're going to *descend*, aren't you? Come down out of that wind and check things out. Put down in some trees at Higbee's or Hidden Valley, maybe catch a yellow-rump to build up your reserves, and wait there until the next morning to see if the wind changes.

"The reason we see so many hawks on west and northwest winds is because they hesitate to cross under those conditions. They come down on north and northeast winds also. North winds are tail winds, which you might think would have to be helpful, but a tired hawk won't cross on a tail wind because he knows if he gets exhausted a third of the way across he can't turn back.

"The wind we *don't* see hawks on is today's wind—an easy, warm, south wind. That's the wind with the most lift. A hawk comes over the Point on a day like today, gliding into a gentle southerly, and conditions are perfect for crossing. He just sets his wings and he's gone. If he's three thousand feet high when he gets out over the water and his glide ratio is eight or ten to one, he's a long way across before he has to start flapping."

Kerlinger first came to Cape May in 1980 to study how sharp-shins crossed the Bay. Equipped with only a pair of binoculars, a hand-held compass, and an anemometer for measuring wind, he spent six weeks at the Point collecting data for his doctoral dissertation. In the fall of 1982 he returned on a post-doctoral fellowship in Clemson University's Avian Migration Mobile Research Center to conduct the study described in Chapter One about the height of the migration over the Point. That research and other radar work led him to a hard-eyed view of the value of traditional hawk counts, especially coastal counts like Cape May's. "Counts using binoculars and spotting scopes are biased toward low-flying birds. If you really wanted to census populations from a hawkwatch, you'd have to use radar.

"Counts at coastal sites like Cape May are also biased toward first-year birds. Nineteen out of every twenty accipiters we see are immatures, making their first trip south. The adult accipiters stay on the ridges because they've learned that's the more direct route, and they're experienced enough to be able to capture prey when they have to. The immies we see end up here because they don't hunt well enough to feed themselves in inland areas. They let the winds lead them to the coast, they follow the shoreline south. For a sharpie or a Cooper's, Cape May is a giant cafeteria. With all the weak and injured migrant songbirds around, they can eat all they want. But how many of those sharpies will make it another thousand miles south to their wintering grounds and then the two or three thousand miles north again in spring? And are their survival rates higher or lower than the immies that follow the ridges? Do different nesting populations tend to follow the coast and other populations the inland route? Or is the coastal route one of two possible routes a hawk will follow depending on the weather he encounters? The only way to answer questions like that is to analyze the banding returns, but the banders haven't quantified their data well enough for anyone to be able to do any analysis of it." Kerlinger rocks back and drums his pen on the desk. "The banders know I feel that way. I have a real love/hate relationship with hawk banders."

After completing his doctorate at the State University of New York at Albany, Kerlinger became an itinerant scholar, working in temporary positions at three universities—Clemson, the University of Calgary, and the University of Southern Mississippi—while he finished his book and published articles on a wide range of ornithological topics, from the habits of snowy owls wintering on the Canadian prairies to the flight strategies of warblers crossing the Gulf of Mexico. Now, in a permanent job he hoped would allow him to focus his research, he has little time for fieldwork. "The best raptor migration in North America is going on right over my head every day, and I'm trapped inside talking on the telephone. I feel like a diabetic in a candy store."

Most of his time is given to the practical problems of keeping the Observatory operating on a bare-bones budget. The 7 × 42 Zeiss binoculars on the windowsill will be his for only one month more. Then he will turn them over to Richard Crossley, the visiting Englishman. "All through June every time Crossley came in here he couldn't keep his eyes off those bins. He kept telling me how much he wished he had a pair of them himself. Finally, I told him, 'OK, Richard, if you do the songbird census for us at Higbee's from July first to October first, I'll give you my Zeisses on the last day.'" Kerlinger shrugs, "What else could I do? We need the data; he wants the bins; and I didn't have anything else to offer him." Zeiss 7 × 42s are widely regarded as the best birding binoculars in existence, but Kerlinger chuckles as he describes the deal, like a garage-sale trader who has pulled off a sly trick. Counting songbirds four or five hours a morning for eighty or ninety mornings, Crossley will put in four hundred hours to earn a $600 pair of binoculars.

CMBO's two dozen hawk banders make even less—most are paid less than a dollar an hour; several receive no pay at all—and this is one reason Kerlinger must be careful when he complains about them not analyzing their data for publication. CMBO bands more hawks each year than any other banding station in the world, and

Kerlinger is frustrated that so few publications have come from that effort, but almost all the banders have been working at the Point much longer than he has, and the data from the banding goes back to the 1960s, before CMBO existed, so Kerlinger's authority to demand analysis is uncertain. And finally, like a new editor trying to energize a blocked writer, he knows that criticism from him is likely to backfire.

For similar reasons, Kerlinger must tread lightly on the question of the accuracy of the hawk count. He doubts the empirical validity of counts in general, and speaking as a raptor biologist, he could criticize them freely. Now, as director of the Cape May Bird Observatory, whose greatest claim to fame is its autumn hawk count, he must be much more careful. Among other problems he runs the risk of alienating Jeff Bouton, who earns as little as the banders, works at least as hard, and is CMBO's most visible representative and spokesman.

Finally, Kerlinger must be careful about criticizing the banding operation or the hawkwatch count because both have long been targets of CMBO's most articulate critic.

On the evening of September 1, Al Nicholson sits in his living room six miles north of the Point, surrounded by landscapes. Framed and unframed oils are stacked upright in two corners of the room; half a dozen hang on each wall. From a distance the paintings look impressionistic. Step within a yard or two, and each becomes a carefully shaped scene—a path under long-limbed oaks or a river bending through woods under huge, imposing skies. The sky is Nicholson's favorite subject. In almost every painting the horizon is low and distant; clouds rise and curl above each scene, radiating light.

Tanned and tall, Nicholson looks fifteen years younger than his sixty years. His hair is dark, his eyebrows are as thick and spiky as caterpillars. He has a soldier's straight frame, large, long-fingered hands, and a voice that wins arguments.

"They say they're doing that platform hawk count for science. It's the very *antithesis* of science. The people on the platform don't move around. They're not interested in studying the conditions or trying to anticipate where the flight will take place. They're not watching at Cox Hall Creek, or Fishing Creek, or even Pond Creek. They're just standing in that one spot at the Point, looking up. If they stand there long enough, they're going to see rarities—there have always been rarities—but where's the discovery? where's the mystery? what are they learning that we don't already know?"

Nicholson has been studying hawks at the Point since, as a boy in the 1930s, he first began visiting on annual trips by train with older birders from Philadelphia. "The night before we came each year I'd be so excited I couldn't sleep. Higbee's was a real wilderness then. You could walk up on eagles standing in the paths."

Nicholson witnessed some of the last of the hawk shoots at the Point on his first trips. Hawks were considered vermin in those not-so-distant days, and during big flights gunners lined Sunset Boulevard to shoot them down as they crossed the road. Roger Tory Peterson, sent by the National Audubon Society to count hawks and observe the shooting, watched eight hundred sharp-shins attempt to cross the road one September morning in 1935. Of that flight, 254 were brought down by gunfire. That evening Peterson, who was lodging in a local home, out of curiosity tasted one of the twenty sharp-shins served at dinner for the family of six, two of whom were gunners. "Like a spy breaking bread with the enemy," he wrote later, "I felt uneasy. I could not tell my hosts I disapproved, for their consciences were clear—weren't they killing the hawks as edible game and at the same time saving all the little songbirds? It would have done no good to explain predation, ecology, and the natural balance to these folks. Having lived at Cape May all their lives, they had a distorted idea of the abundance of hawks."

Nicholson, who couldn't bear to watch the shooting and remembers it today as background gunfire, had the same feeling about the gunners. "They were a pretty motley crew, poor farmers most of them, not educated at all. I was just a boy, but I knew it was no use arguing with them. To them, every raptor was only a chicken hawk. They shot bluejays, flickers, red-tails, ospreys, even great blue herons. They just didn't know any better."

Currently, he is more distressed by what he believes is a widespread misunderstanding of Cape May's hawk flight today by those who should know better. In Nicholson's view the hawk flight was underestimated by local observers before the Audubon-sponsored counts by Peterson and others in the 1930s, and is now grossly overestimated. "In the twenties," he says, "the local birders only went out looking for raptors when a northwest wind blew. They only had little dime-store binoculars, and they weren't very serious. Most of those old-time birders were botanists, and they spent most of the time looking at the ground."

Nicholson isn't sure what to make of the five counts of the 1930s: 14,060 hawks counted in 1931; 10,611 in 1932; 13,452 in 1935; 5,023 in 1936; and 8,377 in 1937. The current counts of fifty to seventy thousand hawks a year seem to him far too high, but the 1930s counts seem too low. "I don't know whether the counters moved around enough, responding to the wind conditions, and I'm certain they missed some birds they should have seen. No one in any of those years reported golden eagle, for instance.

"Even up into the fifties and sixties, people didn't really study raptors at the Point, and no one believed my reports. Ernie Choate did his first season-long count in 1965, and I remember telling him that golden eagles were regular migrants in the fall in Cape May. He just laughed, told me I was crazy, and dismissed it.

Morning, Noon, and Night
❖

"I would watch at Cox Hall Creek above the Canal. There were peregrines at Cox Hall every day from mid-September to mid-October. At low tide, you could watch them going through aerial maneuvers, chasing gulls out over the Bay, doing cartwheels in the sky. At high tide, they just stayed there, perched in the trees. That dune forest was an environment they loved. It's gone now, of course.

"The only other person really studying hawk flights at the Point back in the sixties was Walter Fritton, a fireman from New York City, and he used to come down and spend the day watching the skies. No one else was doing that, not even the birders from the Delaware Valley Ornithological Club, who were pretty serious. They'd stand at the Point for a few hours, then get restless and go off into the woods to look for warblers."

When Nicholson pronounces "warblers," his lips move as if he were trying to spit out a rotten grape.

"Fritton and I used to meet at the Point and watch the flights evolve. First, puffy clouds coming on northwest winds, then peregrine falcons weaving figure eights in the sky, then golden eagles so high they looked like a mirage.

"We learned the subtleties. Fogs and clouds need to clear out during the night for the hawk flight to be good. If a fog lingers into late morning, no hawks. We learned when to move around, too. After a frost in October when it warms up again, and the wind is from the southeast, that's when to go to Town Bank.

"Peregrines are abundant in years when there are gentle winds and lambent, humid conditions and the fall slips away without too much turbulence. The grays and the fall tones deepen. The flickers stay around, and the other birds the peregrines prey on, so the peregrines aren't in a great rush to move, either.

"Later in the season, when the winds shift to west, and the sky shows that vibrating light, that's when to look for goshawks.

"Down at the Point, Fritton and I used to stand under the concrete bunkers that were still there then, part of the Coast Guard Station. Those bunkers were crucial for watching. They protected you from the wind and sun, and that opened up the whole sky. Ernie Choate said they were an eyesore and had them taken down. Now they've got that enormous parking lot there, and the platform, and those crowds, and the mystery is gone."

Nicholson continues to believe the hawk count is flawed because the counters are counting the same hawks again and again. "The counters on the platform tell you that the peregrines are crossing. But how do they know that? They've never watched them from anywhere else. A lot of those peregrines go way out over the Bay, circle back, and return near Town Bank. The counts at the platform are a distortion because they're re-counting the same birds. The hawks get into a cycle. They fly past the Point and out over the water, then come back up the Bayshore, all the way up to Fortescue in Cumberland County. They go into the interior there and sit down. Then on the next northwest wind they do another flight. There are vast duplications on the counts of all those birds, peregrines especially."

Nicholson seldom visits the platform or talks to visitors, so his birding skills and raptor knowledge are still underappreciated by other birders. He is better known for his opposition to the banding project and his decades-long battle with the County Mosquito Commission. "The madman," he is called by some who have never talked to him, "that crazy painter." In fact, he is a thoughtful naturalist, a sharp-eyed observer, and an expert hawkwatcher.

Several of the very best birding areas on the peninsula were unknown to birders before Nicholson discovered them in the 1960s and '70s: Goshen Landing, Jake's Landing, Reed's Beach, the Beanery, the Maurice River. In 1976, Nicholson rediscovered the Mississippi kite in Cape May County. The previous last report of the species had been a sighting by Witmer Stone in 1924 at Higbee's Woods near the old turnpike. "I saw my first at Pond Creek soaring with broad-wings. I'd never seen a kite before, but I had studied Stone's description, and I knew that's what it had to be—the undulating flight, the buoyant delicacy, the lovely head shape. It was the kind of weather we've learned to expect them now—hot and muggy, with a high haze—the 'kite sky.' "

Nicholson found Mississippi kites at the Point in each of the next eight years, and they have been seen every spring since, usually between May 25 and June 1. The regularity of these appearances so far north of their breeding range—they nest from the Carolinas south and west to Oklahoma—suggests that overshooting is a normal consequence of their long migration from Central and South America, where they winter. "I'm certain they've occurred here for years. No one else was looking for them. Everybody was too busy looking for warblers."

October 24, 1978, is the date of Nicholson's most famous find. He was working as the State Park naturalist that fall, and was standing outside the park office next to the Lighthouse when an accipiter passed overhead. "It looked like a sharp-shin, but the breast was too dark and the tail was too long. And the flight was different, too—more like a merlin's wingbeat than a sharp-shin's. I loped out to the hawkwatch to tell people about it."

Clay Sutton and Pete Dunne were at the watch and had studied the hawk as it went by. Both had noticed its dark breast and odd silhouette, though they had each assumed it was a melanistic sharp-shin. The strange wingbeat had led Dunne to wonder if the bird was injured. Then Nicholson trotted up. "Wasn't that a lovely European sparrowhawk?" he asked.

As the hawk turned and came back, and then circled the Lighthouse area for several minutes, Dunne and Sutton became convinced Nicholson's identification was correct, and Sutton photographed the bird. That murky, blurred photograph is still the best evidence that *Accipiter nisus,* the European sparrowhawk, has occurred on this side of the Atlantic Ocean. No other sighting of the species in continental North America has ever been documented.

"I've seen at least two others since," Nicholson says with a shrug, "one just three Novembers ago. No one believed me, of course.

"Pete Dunne says a birder's credibility is like virginity: you can only lose it once. But credibility is more subtle than that. If you don't report what you saw because you're worried about people doubting you or because some pseudo-scientific records committee might say no to you, what's your credibility worth then? You have to follow your instincts—and your instincts have to be sharp.

"Most of the birdwatchers on the platform don't study shape. That's the key to identifying raptors. Most people are only looking for rarities. So they're not going to study a sharp-shin over and over again—or a red-tail over and over again. They don't have the interest or the patience.

"It doesn't matter very much, of course. Everything is pretty much known now, anyway. The mysteries are gone. You used to see a Swainson's hawk and you wouldn't know what it was. That added mystery. You knew it was something different, but that was all.

"Now the banders trap Swainson's hawks every year, and the drive to establish more records has become meaningless. They say they do the banding for research and educational reasons. It has *nothing* to do with research or education. Trapping birds *deadens* research. It *deadens* conservation. Those people aren't learning anything. It's personal amusement. They should devote all that energy they have into something useful. They get fascinated with touching the birds. You see broad-wings without tails at Pond Creek, and marsh hawks without tails. They've lost their tails in the banding operation.

"With the toll taken on hawks, they should forget their sport of banding—or at least acknowledge it *is* a sport, not a science. A golden eagle comes over and all the banders want to do is catch it. They grab it and run it over to another banding station—to show it to their friends. What they're saying is 'This resource is ours. This golden eagle belongs to *us.*'"

Nicholson glowers at the ceiling as he speaks, his hands thrust deep in his pockets. "Birders aren't much better. They're not tuned in to the mystery. They don't want to study the conditions, or learn how to anticipate the flight, or try to understand why the hawks are flying. They play only one game: Name the Bird. Hawkwatching is like bird art today, a yuppie thing to do. Bird art has become nonart because they've taken away the imaginative elements. It's become a rendering skill: they focus on every feather. Which negates education and appreciation. It *sterilizes.* Hawkwatching has been sterilized the same way.

"In America today, a substitute culture has taken over for real culture. Everything is sterilized and commercialized, and people follow the leader like sheep. Most of the conservation organizations are social clubs—selling T-shirts to pay for refreshments at the next meeting. Where's the education and the conservation?

"The reason CMBO is always talking about record hawk counts is because people want hype, the Madison Avenue spin; otherwise, they're not interested and they don't respond. All those people Pete Dunne convinces to come down here for the Audubon weekends—they wouldn't be here if he wasn't so good at PR."

"PR" makes Nicholson's lips spit out the grape again.

"And what do those people do while they're here? They sit at a workshop, watch a slide show, and then follow the leader over to the hawkwatch. Then they stand on the platform for thirty minutes, packed together like cattle, until someone shouts 'Peregrine!' and a bird flies by. And that's it. Then they get back in their cars to drive up the Parkway and home again. A weekend like that doesn't educate; it *sterilizes*. It leads people away from the wellsprings of imagination and appreciation.

"The old-time birders in the thirties and forties carried their lunch in a bag. A lot of them were indigent. They didn't have cars. Some of them didn't have any teeth. But they were *close to it*. I remember one time meeting old Richard Miller at Tinicum when I was a kid. It was the middle of the winter and he was looking for owl droppings. 'Have you seen any excrement?' he asked me. I didn't know what he said. 'Have you seen any *excrement?*' I could hardly understand him because he had no teeth, he talked like he had marbles in his mouth. But his eyes were lit up and excited. He was *close to it*.

"Nowadays birders are programmed by the modern world. They all have two cars, three televisions, and department-store tastes. What does birdwatching mean to them? Zeiss binoculars and Gore-Tex jackets.

"Hawk flights are wonderful, but they've become vulgarized and exploited. The peregrine flight is still good, of course, in the first week of October, but there are too many people around then. I don't even go out hawkwatching anymore until November, when it's too cold and blustery to paint. November is when the red-tails come, anyway, and the golden eagles—and the birders have cleared out and left us in peace."

Chapter Four

❖

DEMO

THE PLATFORM IS CROWDED WITH A SATURDAY MORNING MIX OF BIRD-ers: three white-haired men with canes; two young mothers scanning the sky while their infants sleep in backpacks; an ornithology class from the University of Maine; a middle-aged couple sharing one pair of binoculars and a thermos of coffee; and a dozen of the platform regulars, including Bob Barber and Vince Elia, two of Jeff Bouton's most frequent companions.

Looking happier than he has all season, Bouton walks around the deck offering chocolate-chip cookies from an Entenmann's box to everyone present. For the first time this fall he is wearing hand counters, one on each forefinger, and a two-way radio pokes from his shirt pocket. The radio is his link to the hawk-banding stations, four of which began operating this week: East Station on the edge of the South Cape May Meadows, North Station in the State Park woods, Far North Station on private property across Sunset Boule-vard, and Hidden Valley Station at the Hidden Valley Ranch a mile and a half north of the platform. Each station communicates to the others through the radios, but the hawkwatcher's position on the raised deck gives him a far wider horizon than any of the dozen banders, and so Bouton is the hub of their network, the banders' common link to the sky.

One piece of news has the platform buzzing: the Montclair Hawk Watch in northern New Jersey recorded 17,400 broad-winged hawks in a single day last week, September 16, one of the largest flights ever seen in the eastern United States. Bouton's largest single-day flight so far has been only 1,200 hawks, he hasn't yet totaled 10,000 for the season, and hawkwatchers are naturally competitive, but he shrugs off the obvious comparison. "Seventeen thousand is probably half of all the birds Montclair will have all fall. We're just started. What have they got to look forward to?"

Bouton's spirits have been buoyed by the platform's first rarity. A northern wheatear spent the afternoon of September 11 feeding in the grassy field next to the parking lot, 150 yards from the platform. A thrush that nests on rocky tundra far north of the tree line in Alaska, the Yukon, Baffin Island, and Labrador, the wheatear is among the most sought-after rarities in North America because its nesting grounds are so remote and because, like no other New World songbird, it winters in Africa. Its migration route apparently leads it directly over the Atlantic Ocean from Canada, and so it is seldom reported anywhere in the United States. It has been recorded in New Jersey fewer than twenty times this century.

Bouton has described its discovery to anyone who would listen at least three times a day every day since, and his presentation has become as polished as a tour guide's. "A guy named Paul Rodewald came running over here, all out of breath. 'How often do you see wheatears here?' he asked me. I looked at Clay, Clay looked at Vince, Vince looked at Bob, Bob looked at me. 'There's one in that field right over there,' he said. 'Come on.' We followed him. I figured it was probably a shrike, it couldn't be a wheatear. Then he says, 'I'm just back from Alaska, and I must have seen a thousand wheatears this summer,' and right then he points straight up. *'There it goes now!'* A little thrush-sized bird with a T in its tail. It was no shrike. Bob ran down the beach chasing it, Clay circled around the park office the other way, Vince and I stood over there by the whale's jawbone in front of the office, thinking that was the last we'd see of that bird. About thirty seconds later here came a gray bird with a T in its tail, zipping right past us and back through those cedars on the edge of the field. We tiptoed over and peeked through the branches. There it was, twenty yards away. *Christ,* that scared me to death. What if we flushed it? What if nobody else saw it? But Clay came back in a couple of minutes, Bob came back next, and pretty soon we had about fifteen scopes on that bird. By the end of the afternoon, every birder in South Jersey was here watching it." Bouton puts down the cookie box and wobbles his hand up and down to show how it fed, bobbing in the grass like a robin.

The hawks are coming by in sets this morning: one sharpie, two kestrels, and a broad-wing in one half-minute stretch; then, after ten minutes of empty sky, two sharpies, a Cooper's, and a kestrel; another ten minutes of nothing; then three kestrels and a broad-wing.

"Hey!" shouts Bouton, interrupting the wheatear's walk. "Bird coming right at us with bowed wings . . . could be . . . it is . . . *peregrine!*" It is a female by size, with a big chest, bright white throat, and rippling wings that flick downward with an elegant snap. "Probably a second-year bird just going into adult plumage," says one watcher. "The secondaries look wider than the primaries. The secondaries haven't molted yet." As she reaches the pond in front of the platform, the falcon kinks her wings in jet fighter position and swoops, dropping in a leisurely descent. The swallows feeding over the pond spill out of her way, but the ducks on the pond's surface sit tight. Thirty feet off the water, the falcon pulls out of her dive, swoops back up to cruising height, and resumes her pumping flight out to the Bay.

"Those ducks are veterans," says Bob Barber. "If any of them flew, it would have been all over. They're safer in the water, and they know it."

At 10:00 A.M. sharp, hawk bander Chris Schultz steps up on a picnic table fifty yards from the platform for the first banding demonstration of the season. With erect, muscular shoulders, curly dark red hair, and a thick, rounded Fu Manchu mustache, he looks athletic, proud, and poised. He is wearing a yellow CMBO T-shirt with a print of a peregrine flying past the Cape May Lighthouse. At his feet seven cylindrical cans rest on their sides. A tail of a hawk protrudes from each.

His audience of seventy or eighty people has assembled quickly, most of them coming directly from their cars. Few carry binoculars or even walk close enough to the platform to read the totals board. While the first kettle of the day circles overhead—a dozen broad-winged hawks climbing a thermal—no one in the demo crowd seems to notice. All eyes and cameras are on Schultz as he pulls a kestrel from a can that once held Pringles Potato Chips.

Schultz turns slowly to face in all directions, holding the hawk high, like an Olympic torchbearer showing the flame to the assembled multitude. "This is an American kestrel, a female," he announces in a clear, loud voice. "You can see she has brown wings and lots of lines in her tail. If you look closer, you'll see that we've banded her. Our bands are supplied to us from U.S. Fish and Wildlife Service, and numbered serially. If the hawk is found and the band is returned, we can learn when and where it was banded, even who banded it."

Demo

❖

A pretty brunette in blue jeans and a buttoned white shirt steps up to join Schultz on the table. "This is my sister Laurie who's visiting this weekend and helping with the demo today." Laurie bends to a second can, straightens up with another kestrel held high, and turns in unison with her brother. *Eeeent! Eeeent! Eeeent! Eeeent! Eeeent!* The two hawks squeak at each other, hackles raised, bills snapping—as if each blames the other for their predicament.

"The bird Laurie is holding is a male. You can see he's smaller than the female I'm holding. Most raptors are sexually dimorphic with the female being larger. Male kestrels also have blue wings, and immature males like this one have a single brown band on the tail."

"How old is he?" someone asks.

"Both these birds were born this year, in late spring or early summer, probably in June, so they're only three months old now. They don't fly as families, by the way. The young migrate first, flying ahead of the adults." Schultz lowers his arm and opens his hand. The kestrel hesitates a long moment in his palm, swiveling her head around as if unsure she is free; then she explodes into flight, pumping away hard out past the platform and over the pond. The crowd cheers.

Schultz pulls a Cooper's hawk out of the next can, and several people in the front row fall back, startled. The hawk is screaming—*Ahhh-eeeehhh! Ahhh-eeeehhh! Ahhh-eeeehhh!*—and flapping hard. Schultz turns his face to the side to avoid the wings, but otherwise ignores the commotion. "This is a male Cooper's hawk, and I want you to look at the size of his feet," he shouts over its cries. With his right hand he tugs on the banded leg and points the foot in all directions. The talons are as thick as a serving fork's prongs. "Many hawks have gotten a reputation as 'chicken hawks.'" *Ahhh-eeeehhh! Ahhh-eeeehhh! Ahhh-eeeehhh!* "Most don't deserve that reputation— they eat insects or snakes or rodents." *Ahhh-eeeehhh! Ahhh-eeeehhh! Ahhh-eeeehhh!* "But the Cooper's is one hawk that *does* deserve that

"Banding allows you to know birds personally."

reputation." *Ahhh-eeeehhh! Ahhh-eeeehhh! Ahhh-eeeehhh!* "It's one hawk that will take a chicken."

"Would it feed on a kestrel?"

Schultz laughs. Head feathers raised, bill gaping fiercely, the Cooper's has turned to scream in the face of the kestrel, *Ahhh-eeeehhh! Ahhh-eeeehhh! Ahhh-eeeehhh!,* which is now silent and cowering—shoulders high, head tucked low—in Laurie Schultz's hand.

After a moment, Laurie Schultz steps back and opens her palm. The male kestrel pumps away; the crowd cheers and claps. Laurie bends to another can and holds up another accipiter, brown cap and spotted breast. "This is a female sharp-shinned hawk Laurie has now, and we'll hold these two close together because there's a myth that the sizes overlap between these two species." The Cooper's screams and snaps at the sharpie; the sharpie, at least ten percent smaller, raises its crest feathers and snaps back. "People say that you can't tell big female sharpies from small male Coopers. We've banded thirty-eight thousand accipiters in more than twenty years of banding here at the Point, and we have yet to find a single example of true overlap. So the two sizes might overlap but, if so, it's a rare event."

"What are those feathers falling off on your hand?"

"Those are down feathers from his breast. He has thirteen thousand feathers, so he won't miss two or three."

Schultz opens his hand and the Cooper's strokes away, up around the pond and then circling back, heading toward the Lighthouse. The crowd claps, whistles, and cheers.

"Will he go across the Bay?"

"If the winds are advantageous, he might. But we know from putting radio transmitters on them that many of the hawks get out about a mile or two and say, 'This is not for me' and swing back. Then they go up the river past Higbee's Beach to find a narrower place to cross.

"What do you think the mortality rate is on these birds?" Schultz asks the crowd. No one answers. "It's *fifty* to *seventy-five* percent," he says. "So chances are that bird will be dead before next March."

The crowd moans.

"It's not the banding. It's nature's way. Great horned owls eat them, and they have other problems. Once a hawk makes it through the first six months, the mortality rate goes way down."

Up on the platform, Jeff Bouton's radio crackles. It's a bander in the North Station: "Warn Schultzie: the last bird, the second harrier, isn't wearing a band. We forgot to put it on."

Bouton hurries down the ramp. The crowd around Schultz has grown to more than a hundred people, however; people in the back are standing on tiptoes, and others are hurrying toward the group from every corner of the parking lot. Bouton can't push through. He circles around the outside trying to catch Schultz's eye from long distance.

Schultz pauses before reaching for his next bird. "I want to tell a story about another hawk that eats birds and about the importance of returning bands," he says. "This time the 'chicken hawk' was a goshawk we banded here at Cape May. The man who found it in North Carolina sent back the band with a letter to us. 'I found your bird in my chicken coop, eating his twenty-ninth chicken.' He shot it. Would we prosecute? No, we wouldn't. It *is* illegal to shoot hawks, but we wouldn't prosecute someone who returns a band. We need that information. The return rate on banded hawks is less than three percent. We have to band an awful lot of birds to get any information, and we're grateful for every single band we get back."

Schultz's enchantment with raptor banding began at a talk given by Bill Clark, Cape May's first raptor bander, at the Vermont Institute of Natural Sciences in 1979. By the time the talk ended, Schultz knew he was hooked, and he pushed his way through the crowd to ask Clark the question Schultz himself is now asked regularly: "How can I get involved working with you at Cape May?" Clark gave the answer he gives all novices, the answer Schultz himself now uses: "I'm glad you're interested, but the blinds at the Point are pretty much full." This is the standard bander-to-novice rejection—with the "pretty much" acting as the qualifier, as the crack in the door. The Point's banders are often accused of running a secret society that is hostile to newcomers. But trapping hawks requires solitude and silence. If every novice who expressed interest in the operation were invited into the blinds, the banders would do nothing all season but entertain human visitors. For simple, practical purposes, all newcom-

ers must be turned away at first request, and all but the most determined and persistent must be denied admission entirely.

Back home, leafing through the handouts Clark had distributed, Schultz found a notice from CMBO announcing an Expanded Hawk Watch, the experimental multiple-site hawk census
planned by Pete Dunne for fall 1979. Schultz drove to the Point that
September, counted hawks all fall from the humdrum lookout he was
assigned on the Cape May Canal, and eventually befriended two
banders, Bob Dittrick and Linda Ellis, who invited him to their blind.
There Schultz met Clark for the second time. Clark hardly batted an
eye. "You got here after all, huh?"

Schultz has returned every fall since and, after trapping more
than five thousand hawks in eight years, he is generally regarded as
the Point's most skillful and expert bander, which means he is one
of the very best in the country. He has also become a full-time
"PRB," as he calls himself, a Professional Raptor Bum.

His year begins in Colorado in April, where he works under
contract for the Colorado Division of Wildlife monitoring the
breeding success of peregrine falcons in the state. Forty pairs of
peregrines nest in Colorado, and all the aeries, like most peregrine
aeries around the world, are on high ledges on cliff faces and moun-
tainsides. Schultz has had to train himself as a rock climber so he can
investigate and band at the sites. Some sites must be approached by
river raft; others only by helicopter. Sometimes the rock faces are so
sheer that he must tie his rope to the struts of the helicopter and then
climb down, dangling over thousands of feet of vertical rock. Occasionally he is defeated. At a site in the Rockies, ten thousand feet up,
the highest peregrine aerie in Colorado, Schultz found it impossible
to pinpoint the peregrines' nesting ledge on the cliff face. The two
falcons, wary of the local golden eagles (which feast on falcon chicks
and are the peregrine's greatest nonhuman enemy), approached their

ledge with such speed and stealth that Schultz and his banding companions couldn't spot the aerie through their binoculars, despite hours
of watching. Eventually, Schultz unfurled a kite painted to look like
a golden eagle in the hope the falcons would fly out and attack, but
the lure didn't work and the aerie remained unvisited, the young
unbanded.

When he can reach an aerie, Schultz collects any eggshell
fragments still on the ledge for laboratory analysis. The enormous and
tragic decline of the peregrine falcon in the United States between
1947 and 1970 was, we know now, primarily the result of DDT.
Living high on the food chain, feeding on birds who themselves are
living high on the food chain, peregrines are particularly sensitive to
pesticide poisoning. The most obvious symptom of DDT toxicity is
eggshell thinning, which leads to addled and broken eggs and nest
failure. By 1969, when the use of DDT was finally restricted in the
United States, it was too late for eastern peregrines. The pre–World
War II population of approximately two thousand pairs nesting east
of the Mississippi had been reduced to zero; today, there are still no
naturally occurring peregrines breeding in the eastern U.S., and there
may never be again. The western population barely survived the
DDT years—the estimated thousand pairs were reduced to a hundred—but with state and federally sponsored protection, their numbers have been slowly increasing recently. Traces of DDE, the natural
breakdown product of DDT, can still be found in the eggs of western
peregrines, probably because many of the swifts and songbirds they
prey upon spend their winters in Central and South America, where
DDT (produced and exported by the United States) is still used
freely.

In a new Colorado program designed to reduce fledgling
mortality to a minimum, Schultz climbs to a peregrine aerie and
removes the eggs from the nest, replacing them with ceramic substi-

tutes to maintain the adults' interest in brooding. Back at the lab, the eggs are kept in incubators until hatching, and the young are fed by human hands for the first week and a half of life, a critically danger-ous period in the wild. On the tenth day Schultz carries the young by helicopter back to the site, rappels to the aerie again, and returns them to their parents. By this point the young peregrines are much larger than newly hatched chicks would be—fluffy and white, all belly and feet, they look like the "Kick Me" Schmoos in *Li'l Abner*—but the adults are apparently undaunted by the instant transformation from eggs to full-throated fledglings. They generally start feeding their young as soon as Schultz climbs off the cliff face.

Each July, as the Colorado falcons are leaving their nests, Schultz goes north to Greenland, where the peregrine nesting season has just begun. He travels first by an Air Force C141 transport plane to Thule, then by prop plane to various gravel bars along the west-central coast of Greenland, and finally by foot, carrying a ninety-pound backpack stuffed with climbing and banding equipment and ten days' supplies. He and his banding partners follow musk ox trails, scanning the rocky cliff sides for a certain *Caloplaca* lichen that grows in falcon excrement and so reveals the peregrines' aerie.

Although the Greenland cliffs are not as steep as the Colorado mountains, the climbing is just as dangerous because the banders have only minimal equipment (long ropes would be too heavy to carry), there is much loose rock, and any serious injury would be a disaster. The banders are alone in one of the most remote areas on Earth. The continental ice sheet towers in the distance. The work is slow. Each aerie is separated from the next by miles of treeless, windswept landscape. Mosquitos whine and bite twenty-four hours a day, except during the fiercest winds, when it sometimes snows.

A hundred nestlings banded in a summer is the most the banders can hope for. Since the return rate on nestlings is even lower than the rate for fledged birds (because nest mortality is so high), Schultz and his partners must consider themselves lucky when a single band is returned from their entire summer's labor.

As poor as the band-return rate is, the survey of nest sites also serves the general purpose of monitoring the falcons' numbers. The peregrines that nest in Greenland are a different race of peregrine, the arctic or tundra race, *Falco peregrinus tundrius,* birds which are visually distinct from the continental race found in Colorado because they are slightly smaller, have thinner mustache markings, and, in their immature plumage, wear blond caps. These were the peregrines that best survived the DDT years, but they too experienced a population crash in the 1950s and 1960s, and traces of DDE are still found in their eggshells. The Greenland surveys are one of the few direct means of monitoring their health. Although the arctic is still relatively pesticide-free, *tundrius* peregrines have one of the longest migrations of any raptor in the world, from the top of the world to the Southern Hemisphere, and they are apparently now being contaminated by pesticides found in their prey *en route* and on their wintering grounds in the Caribbean and South America.

Leaving Greenland in August, Schultz returns briefly to Colorado, then packs for New Jersey. In recent years he has taken over the day-to-day directorship of the Point's banding operation from Clark, who has become a kind of director emeritus. Schultz arrives about Labor Day to organize "Set Up," the two-week operation all but the most dedicated banders avoid—clearing the undergrowth, rebuilding the blinds, repairing the traps, stringing the nets, sheltering and feeding the lure birds. The hawk banding starts in mid-September, as Cape May's daily flights begin to peak.

Schultz's favorite of the Point's five banding stations is East, the station where peregrines are trapped most frequently, and he dreams of recapturing at the southern tip of New Jersey a bird he has banded above the Arctic Circle in Greenland, three thousand miles north.

As Schultz pulls out his sixth bird, a harrier, someone asks, "How do you capture them?" It is a question asked at most demos, and as always, Schultz ducks it.

"We have thirty different traps at five stations along the flight lanes," he says quickly. "We use mist nets and other techniques. It doesn't matter. What matters is the band. We put the birds in these cans during the banding process. That quiets them and calms them down. The processing usually takes about ten minutes. Last year we caught the same peregrine falcon three times in a couple of hours, then she was recaptured on Assateague Island in Virginia a couple of days after that. Now tell me that bird was stressed by banding."

The questioner has not mentioned stress, of course, but Schultz, like all the banders, is wary of revealing any details of the trapping techniques, especially that all the banding stations use spring-triggered bow traps in addition to the passive and far less dangerous mist nets, and that all use live lures—starlings, house sparrows, and pigeons. He also knows that up on the platform, where at least half the birders have turned to listen to his presentation, there are people who believe banding *does* stress the hawks.

Schultz is particularly sensitive to such criticism since the demo several years ago when, with a hundred people looking on, he pulled a limp Cooper's hawk out of a can. It was dead, apparently having had a seizure or suffocated in its confinement. After a long pause, Schultz regained his composure, turned back to the crowd, and explained that every once in a while a hawk died in the banding operation. Even when all possible precautions were taken and banders were as gentle and caring as they could be, he said, something might still go wrong sometimes. "There are inevitable accidents," he said.

The people watching seemed to accept this explanation. Then, in a churchlike silence, Schultz reached for his next can, and pulled out another dead Cooper's.

Those who witnessed that demo do not seem to remember what Schultz said next. What they remember is he somehow held his poise: he looked his listeners in the eyes, his voice remained strong. As the demo ended, several listeners came up to shake his hand, thank him for the honest talk, and express their condolences. After the last of them was gone, Schultz broke down and wept.

"How many hawks have you killed today?!" people have shouted out at his demos in the years since. One Saturday, after Schultz explained that the return rate on bands was about three percent, a man at the back of the crowd shouted, "How do you know you haven't killed the other ninety-seven percent?" As Schultz tried to answer, the man shouted him down, "How do you know you haven't killed the rest!? . . . How do you *know!?* . . . How do you *know!?*"

"Humane-iacs," the banders call these people, or "bunny huggers." But they are not a homogeneous group. Some seem to be simply passersby, annoyed because they have happened upon a scene where a wild animal is being restrained. A few are animal-rights activists. Others are well-informed naturalists including a few, like Al Nicholson, who have been studying migration at the Point longer than any of the banders. The man who shouted Schultz down has been a professional wilderness tour leader and president of a local ornithological club. These people generally share Nicholson's views that the banding is a worthless and dangerous intrusion in the lives of birds, lives that are already endangered enough by human activity. "What has the banding proved after all these years?" cynics have asked Schultz. "That birds fly north in spring and south in fall?"

Banding, or "ringing," as it is called in England, has indeed proved birds fly north in spring and south in fall—and much else too. In fact, almost everything we know with certainty and precision about bird migration has been discovered or proven through banding, and it is not an exaggeration to say that the numbered leg band has been as important a tool in the science of avian migration as the telescope has been in planetary astronomy.

Migration is so hard to study by direct observation and general impression that Aristotle's claim that certain species transmogrified (the redstarts of summer became the robins of winter; garden warblers became blackcaps) was believed into the seventeenth century, and his theory that swallows and other species overwintered by hibernation was accepted well into the eighteenth century by many scientists, including Carolus Linnaeus, the great Swedish naturalist who originated the system of taxonomic classification and lived until 1778. Studying and documenting migration was difficult because few observers had traveled far enough to study birds on both their summer and winter grounds and because no satisfactory method of identifying individual birds to trace their seasonal movements had been developed. Some early researchers tried marking birds by painting or dying their feathers, but most migrant species molt their feathers just before or just after migration (and all birds molt all their feathers at least once annually), so the method was useless for long-term investigations. Coastal observers watching swallows appear in spring from out over the sea and disappear over the horizon in fall came to the conclusion that the birds hibernated underwater. A woodcut from the sixteenth century shows fishermen pulling up a net filled equally with fish and swallows. An alternative theory was that birds flew straight up into the heavens each fall. The anonymous author of a pamphlet published in England in 1703 argued that an overwater crossing of the ocean or the sea would be an unthinkably taxing journey for any bird and suggested that birds' wintering ground was the moon.

The first to attempt to use leg markings to investigate migration was apparently an unknown birdwatcher in the Ottoman Em-
pire. In 1710 a gray heron (*Ardea cinerea*, the European equivalent
of our great blue heron) was captured in Germany with a metal ring
on its leg. The ring indicated the bird had come from Turkey, but
the German captors had no way to identify the bander or to notify
him that his bird had been found. The first European to mark birds
by banding was the scholar and entomologist Johann Leonhard
Frisch, director of historical studies at the Berlin Academy of
Sciences, who tied silver strings to the legs of swallows in 1740. A
century later John James Audubon reported in his *The Birds of America* that he had tied silver wires to the legs of fledgling phoebes at
Mill Grove Farm in eastern Pennsylvania one summer and had found
that two of them returned to the area the following spring. This is
one of the earliest documented examples proving migrant birds could
find their way back to the same breeding grounds in subsequent years.

Technical problems limited the value of bird banding
throughout the nineteenth century. No satisfactory banding material
was available—metal bands were heavy and interfered with flight;
strings, wires, and leather loops fell off or disintegrated—and, because
the only observer likely to recognize his own band was the bander
himself, the odds against successful recoveries were enormous. The
gaps in our knowledge of migration were still glaring and countless
at the start of the twentieth century. Ornithologists could not answer
the most fundamental questions about the migration patterns and
orientation abilities of birds. Did they follow the same migration
routes each year? Could the same individuals find their way back to
the same wintering sites? Was a bird's understanding of its migration
route instinctive or learned? Did mated pairs migrate together? Did
families migrate together? The migration routes of such common
species as the arctic tern and the chimney swift were mysteries. The

arctic tern nested by the tens of thousands up the coast of New England and throughout arctic Canada, but individuals were seldom seen south of Cape Cod along either the North American or South American coast. Where did all those birds go? The same question was asked about the chimney swift, one of the most numerous birds on the continent. Hundreds of thousands of swifts could be seen flying south each fall through Texas and Mexico and coming north through Texas and the Gulf states each spring, but where they spent the months of November to March could not be established. No wintering localities had ever been found.

In 1899 a Danish schoolmaster, Hans Mortensen, pushed banding into the forefront of migration research when he devised the first aluminum leg bands and used them in a successful study of the migration and nesting habits of white storks, common teal, and starlings. His bands were so light that they imposed no burden on the bird's flying abilities, yet they were strong enough to last the bird's lifetime. The next technical refinement came two years later and was developed simultaneously and apparently independently by Paul Bartsch of the Smithsonian Museum in Washington and by the German Ornithological Society in Rossitten, Germany. Both used aluminum bands that were numbered serially and marked to identify their source. Bartsch banded 100 black-crowned night herons with numbered bands carrying the legend RETURN TO SMITHSONIAN. The Germans banded gulls, hawks, and other birds with numbered bands and a one-word legend: ROSSITTEN. Some uncertainty followed: one of the first banded gulls found in France was believed to have come from a shipwrecked sailor from a boat named *Rossitten*; an eagle shot in Bulgaria wearing a band with the number 1285 was thought to be 600 years old. Soon thereafter, however, banding caught the imagination of professionals and amateurs alike, and dozens of trapping stations were established throughout Europe and North America. In August of 1917, an arctic tern that had been banded on Eastern Egg Rock, Maine, in July of 1913 was found dead on the Niger River Delta in West Africa. Hardly a season has passed since when the recovery of one banded bird or another—or a series of recoveries— has not startled the ornithological community with some new piece of evidence about the complexity of avian migration.

Today, thanks to banding recoveries, we know migration is no simple north-south movement and that each migrating species follows its own particular route. The arctic tern, the planet's longest traveler, flies east from New England, crossing the Atlantic to Europe and Africa, and then goes south across the Antarctic Circle to winter in the circumpolar regions at the bottom of the world. Other species follow shorter but equally precise routes and return each year to the same wintering grounds. In the 1920s the ornithologist S. P. Baldwin trapped white-throated sparrows returning to the same bush at his banding station in Thomasville, Georgia, for several years running. This *winter-site fidelity* has since been proven to be the rule for hundreds of other species and tens of thousands of individual birds. We know now that golden plovers that nest in Alaska stop on their journey south in Hawaii, then continue onward to winter on the Marquesas Islands, that the arctic warblers of Alaska winter in the Philippines, that scarlet tanagers move south from the breeding grounds in the northeastern United States to winter in Colombia and Bolivia, that the same individual black-and-white warblers will re- turn year after year (apparently as long as they live) to the same couple of acres of woodland in Jamaica, and that the same individual wood thrushes return year after year to the same patch of habitat in Belize. We also know now, since banding recoveries solved the last mysteries, where all North American species summer and winter. The mystery of the chimney swift's wintering territory was finally solved in 1944 when an Indian hunter returned the bands of thirteen swifts he had shot in the Amazonian jungle of eastern Peru.

Demo

Banding has led to other discoveries also. Recoveries of banded black-crowned night herons, the first species banded in numbers in North America, demonstrated that the birds flew *north* after fledging. This *post-breeding dispersal* of young birds has since been shown to be far more common than researchers suspected before banding made tracing the phenomenon possible. In the 1940s, in what became an internationally famous operation, Charles Broley, a retired banker, climbed pine trees to band bald eagles in the woods around Tampa and startled migration scientists by discovering that Florida's eagles, both adults and young, perform a reverse migration each year, migrating north after their nesting season to New England and Canada. It was a phenomenon no one had predicted or apparently even suspected.

"Banding allows us to know birds *personally*," the biologist and veteran bander John Kricher has explained. "It enables us to study a bird not only as a member of its species but also as a single animal with an individual identity. A band is a bird's dog tag. Will that robin nesting in that tree there come back and nest in the same tree next year? Will it have the same mate? And if both come back next year, will they come back the year after that? How many different mates will a robin have in its lifetime? How long does a robin live? The only way we can answer questions like those is through banding."

Because of banding, we know today that most birds are much less faithful to their mates than previously believed: since trapping and banding have enabled researchers to identify the sex of birds and to trace their movements, polygamous pairings have proven to be a rule among many species. We know too that only a handful of species migrate in family groups. Most mates separate at the nesting grounds, migrate south on different routes, and leave their young behind to find their own way south. Our knowledge of the life expectancies of wild birds comes entirely from banding records. No simpler or more reliable method of measuring age has been devised. In 1939 a

Fastening a band on a kestrel

black-headed gull was recovered in central Europe with a band reported to be as thin as paper. The bird had been banded in 1914, twenty-five years earlier, in one of the earliest banding operations in Europe. Longer life spans have since been established—thirty-four years for an arctic tern, thirty years for a great frigatebird, twenty-seven years for a western gull—and the maximum life expectancy of the Laysan albatross is still unknown because banding is such a relatively new methodology. Hundreds of the first Laysan albatrosses banded in the 1950s are still alive, and still wearing their original bands.

Demo

❖

Today two thousand licensed banders band about a million birds annually in the United States and Canada. About sixty thousand banded birds are retrapped or found dead each year. The Cape May hawk-banding project, which Bill Clark began by himself in 1966, has continued now for more than twenty years. More than seventy-five thousand hawks have been trapped here. In recent years, with four or five stations in operation seven days a week and two dozen banders sharing duties and responsibilities from mid-September to late November, CMBO has been capturing between four and five thousand hawks a season, more than any other hawk-banding project in North America.

"How many birds captured here have been banded somewhere else?" a spectator asks Schultz.

"We get ten or twenty each year," says Schultz. *"Foreign recoveries*, we call them."

"How many birds banded here are recaptured here the next year?" asks someone else.

"About two or three a year, out of five thousand banded. It seems hawks learn that Cape May is a mistake. We get more returns on birds banded here caught inland on the ridges in subsequent years. Those birds have learned that the coastal route leads to the Point, and on their next migration they stay inland and follow the ridges south. They don't come back to Cape May."

"How does a hawk catch his food—with his bill or his feet?"

"Watch this," says Schultz. He tugs on the harrier's leg, and the leg extends four inches. The man who asked the question stands on his tiptoes to see. Schultz tugs more, and the leg extends another two inches. Several people whistle; others shake their heads in amazement. Schultz tugs more, then holds the bird up high so all can see. The hawk's leg seems longer than the hawk, and people are laughing, poking each other, and pointing. "If you've ever seen harriers zigzagging over a marsh," says Schultz, "now you know what they're doing. They're reaching down into the phragmites to grab the rodents they prey on."

Bouton is still circling the outside of the crowd trying to signal to Schultz about the unbanded harrier when Schultz pulls it out of his last can. Schultz's look of disappointment, one downcast eyebrow, is gone in an instant. "Here's a bird we've made a mistake on," he announces with a smile. "Can anyone see the problem?"

"No band," says someone in the front.

"Right. We caught this bird just before the demo was due to start, and it looks like in our hurry to be able to show him to you we forgot the band. Now, am I going to take this one back to the station to put the band on that should be there? No, I'm not. This bird has had enough of us today. I'm going to let him go. You've all heard fisherman's stories about the one that got away. Now here's a bander's story." He releases the bird and it flaps away hard. The crowd cheers and applauds in approval.

The audience disperses almost as quickly as the released hawks. Four people cluster around Schultz to ask a few last questions; three others shuffle up the ramp to the platform, then stand shyly in the corner as if worried someone will ask for their membership card. All the rest go immediately back to their cars, climb in, and drive off, most without a single look skyward. Ten minutes after the second harrier headed out over the Bay, the parking lot is three-quarters empty.

When CMBO made the banding demos a regular weekend feature at the Point a few years ago, some banders feared (and some antibanders hoped) that it would be the end of the banding project. The thought was that the beginner birders it was believed the demos would attract would not understand banding, nor would they be interested in the esoterica of migration science. Once they saw what was going on, they would side with the "humane-iacs" and grow angry at the idea of birds' being caught to have their legs encumbered by aluminum bracelets. What has happened is something unexpected. The demos have brought a new audience to the Point, people who are not birders, who don't even carry binoculars—who are intrigued by hawks but not by birdwatching. "The hawk-banding demonstration," Peter Dunne once argued, "reaches a segment of the population that is not reached by any other conservation mechanism. *None.*" The regular size of Schultz's crowds, up to two hundred people four times a weekend, suggests too that this audience probably outnumbers the birding crowd by at least an order of magnitude. Birding from the platform requires a certain mind-set, most of all a willingness to be satisfied with murky, distant looks at the hawks. You can't see a harrier's leg in full extension from the platform, or hear a Cooper's scream, or show your child the erected feathers on the crest of a kestrel. Schultz gives his audience all of this and more, and among the general public, residents and visitors who are not interested in squinting at birds or puzzling over identifications, CMBO's banding demos are its best offering, its most effective outreach program to potential members and supporters of conservation legislation.

About noon an impromptu demo begins, this one conducted by a young man standing waist-deep in the small plot of goldenrod and asters to the right of the platform. "I've had eleven species of butterflies right in this patch since the sun broke through," he tells the birders standing above him at the railing, then reads from his notebook: "Gray hairstreak, eastern tailed blue, white-M hairstreak, buckeye, orange sulphur, cabbage white, variegated fritillary, pearl crescent, American painted lady, sachem, and least skipper." He is Nick Wagerik from New York City. A magnifying glass hangs from a string looped around his binoculars. He steps slowly from one plant to the next, touching the stalks below the white and gold flowers as carefully as a hothouse gardener inspecting orchids.

"What's that small one there?" asks a hawkwatcher from the platform. "Sort of rust colored."

"That's a least skipper. . . . There's an American painted lady. . . . There's a pair of least skippers mating. . . . There's a buckeye. . . . There's an ailanthus webworm moth."

"Moths fly in daytime?"

"Oh, yeah. There are so many moths; lots of them are diurnal. Birders think identifying fall warblers is hard. Just look at the Peterson moth book sometime.

"Here's a gray hairstreak," Wagerik says, pointing to a butterfly the size of a nickel with a dot of orange flame at its rear end the size of a match head. "See the little tails on the hindwings? They're not really tails. It's thought they mimic antennae, to draw predators to the wrong end. See how it's rubbing them back and forth? You find them sometimes with those little false tails missing. The bird went for the wrong end, and the butterfly flew away in the other direction."

A dozen birders are now leaning over the railing, several struggling to aim their binoculars where Wagerik points. "Last June when the birds were slow," says one woman, "I thought I'd try to learn the flowers and insects."

"*Good* for you. You'll never concentrate on birds the same way. Birds got me started too. But now I'm really into insects, not just butterflies and moths. Dragonflies, ants, wasps. I love wasps."

"But where do the butterflies go when the sun isn't out?"

Wagerik laughs. "They disappear. . . . *Look*! Here's a gray hairstreak with part of its hindwing eaten off! That's what I was talking about. A bird probably got him."

At this, four birders hurry down the ramp to join Wagerik

in the weeds. He hands the first his magnifying glass, and she bends close to look, then passes the lens to the birder behind her.

"There are just about as many butterflies in North America as birds," says Wagerik, addressing the birders still on the platform. "Seven hundred and fifty species or so. Butterflies migrate too, and so do a lot of other insects—the green darner, the heroic darner. In England they get monarch butterflies that have crossed the Atlantic— here's a white-M hairstreak!"

A middle-aged man in a Phillies cap high-steps over quickly, takes the magnifying glass Wagerik hands him, turns his cap around backward, and bends close. The M is a chalk-white pin scratch on the underside of the insect's tiny gray wing. "Lovely!" he says. "I guess that's a lifer for me."

Others climb down to lean over. Wagerik loops his magnifying glass off his binoculars and passes it around. Ten birders have now joined him in the patch, wading through the weeds, crouching to turn over leaves, and pointing to butterflies as they swirl into the air, flashing their colors like flakes of confetti.

"Cape May is just as famous among butterfly watchers as it is among birders," Wagerik tells them. "Southerly breezes blow southern butterflies up here just as they bring up southern birds after they've finished nesting. But you need the right winds, and it's like birds: you can never predict it."

He looks over his shoulder in the direction of Delaware. "I'm still waiting for the big push."

An hour later, an eagle appears—in "Eagle Corner," the area northeast of the platform in the direction of Far North Banding Station beyond Sunset Boulevard. It's an immature bald eagle, with a dark head and much white under its enormous wings, and it flies straight toward the platform, with a downstroke as heavy as the flap of a rug.

Finally, directly above the platform, the eagle holds it wings outspread and soars upward, riding higher and higher.

"It's a White Belly II," Jeff Bouton points out. Unlike almost all other North American birds, which lose and replace all their feathers once or twice a year, bald eagles molt their feathers year-round. They also need five years to reach adulthood, one of the longest periods of North American birds, and hawkwatchers who have memorized the sequences of feather replacement can age an eagle at five hundred yards. A White Belly II is a bird in its third year of life.

Half an hour after the eagle heads out over the Bay, three pigeons come flapping at top speed from the Bunker heading across the pond. A Cooper's hawk is closing on them from behind. Two pigeons curl right and away as they reach the pavilion roof, the third curls left and down, and the Cooper's catches it and drives it to the ground. They hit with a thump and a squawk forty yards from the platform, five yards from a man at the pavilion's trash basket. The man freezes, empty soda can in hand, while the Cooper's stands on the pigeon and tilts its head sideways to look at him. One long glance and the hawk is gone. The pigeon, red eyes wide, squats in place for a moment longer, tilting its head back and forth to study the sky, then flies away—apparently unhurt. The man drops the soda can in the basket. The birders on the platform applaud.

"My favorite platform stoop," says Bob Barber, "was that merlin a couple of years ago that came over the pond, then zoomed right into the phrags and came up with a yellow-billed cuckoo. Then it flew off toward Delaware carrying the cuckoo, which was squealing all the way."

"Remember the merlin on the house sparrow on the sidewalk last year?" Bouton asks. "Remember that explosion of feathers?"

"Yeah, that was great," says Elia. "In a good stoop you have to have an explosion of feathers."

Demo

❖

"Hey, out over the ocean, what are those?" someone calls out. Two long-billed, medium-large shorebirds are pumping east, flying away, fifty yards offshore and three hundred yards from the platform at the closest point. Everyone turns to watch, but then there is a very rare moment on the platform—a long silence. No one knows what they are.

"They were godwit-sized," says Bouton after the birds have dropped below the waves on the horizon.

"The one on the left was definitely dark," says Elia.

"I thought greater yellowlegs at first," says Barber, "but the wingbeat didn't look right."

"They were godwit-sized."

"They weren't knots, were they?"

"No way! Did you see those bills?"

"They were godwit-sized."

"Jeff! Jeff!" says a newcomer pushing into the circle to point Bouton in the opposite direction. "What are these?"

Bouton lifts his bins, "A-10s. Three A-10s coming out of Dover Air Force Base."

The newcomer flinches as if he's been bopped on the head with a chalkboard pointer. He looks again. "Their wings are flat, just like eagles."

"Uh-huh. Maybe a little too flat."

By 2:30, when Pat Sutton stops to eat lunch on the platform, the sky has grown cloudy. A thin ash-blond in her mid-thirties, Sutton is CMBO's teacher naturalist, but like Paul Kerlinger, she spends most days trapped indoors, burdened by her office work. She is seldom seen at the Point by herself. Ordinarily, her only appearances are while she is leading a field trip, when she is in what she calls her "leader mode."

Pat claims she learned her teaching method from her husband,

Clay, on her very first birding trip, when he took her to Florida with him twelve years ago. "I must have asked him a hundred and fifty times a day, 'What's that one?' and 'What's that one?' He'd never been there before and he was hoping to see the Florida specialities—caracara, short-tailed hawk, burrowing owl—but I didn't know what a snowy egret looked like. I had never looked through a pair of binoculars or a telescope ever in my life. That didn't matter to him, though. Not one time did he tell me, 'Oh, that's only a such and such.' He was just as excited about the everyday birds as I was, or at least he pretended to be. That was what made me a birder, that *enthusiasm.* I'll never forget that. It changed my whole way of looking at the world."

Clay discounts this history, however, and will take no responsibility for Pat's "leader mode." He likes to tell about the time several years ago when he was standing on the platform in a snowstorm as a long line of birders came walking out of the woods led by a woman in a yellow slicker who seemed to be dancing from one end of the line to the other, wheeling and pointing, shouting out the names of birds in between little hops of excitement. "Who *is* that woman?" Clay had time to think—before lifting his binoculars and realizing he was looking at his own wife.

"Al Nicholson always says birders are too hung up on rarities," says Pat, "and I agree with him one hundred percent. That's one of the reasons a lot of the best birders make such piss-poor leaders. I wish I had a penny for every time I've heard a leader at the head of a line of beginners announce, 'Well, folks, there's *nothing* around here today.' That just kills a trip. Why say something like that? People coming out on a field trip who haven't been birding much before don't know what's exciting. They're taking their cues from you. They're happy getting their first good look at a black-and-white warbler. If you're excited, they're excited. And if you're not excited, why are you leading the trip? What's the purpose?"

Demo

❖

Sutton's skills are sorely tested every Saturday night when she leads CMBO's owl walks. The walks are her idea, for owls are her favorite birds, but a noisy group of beginning birders marching into the dark has nearly as much chance of coming upon a roosting owl as they do of spotting the next comet. Sutton's groups have grown steadily larger each week, but they rarely even hear an owl. After a long, birdless walk in the cold, Sutton is generally forced to arrange the group in a circle under the Lighthouse and have them stare upward into the whirl of the light in the hope of spotting the shadow of an owl passing through. When they grow tired of that, Sutton does owl calls for them, until the cold and the dark force the last survivors back to their cars. The next week many of them come again, bringing their friends, to give it another try. "I don't understand why they keep coming, and I really feel kind of guilty about it," she says. "The chances of seeing something are so low. But every week they're here again. So I keep trying." She shrugs. "You won't see anything unless you're out there looking, right?"

"You ought to try running one of those trips before dawn some Sunday," says Vince Elia. "I'll come. I think the odds are better in the morning."

"*Yeah!*" says Pat. "Let's do it."

"We'll wait until the next full moon and then have everybody stand on the beach," Elia says. "The light reflects off the sand, and you can see them better, barn owls especially."

"*Yeah!*" says Sutton. "I've seen them that way." Then she gives her imitation of the barn owl's spitting cry, "*Skschhhhhhhhhh!*" like a cat meeting a Doberman in a dead-end alley. Twenty birders jump, spin, and stare.

By 3:30 the cloud cover is complete. An occasional sharpie or Cooper's passes, but the most visible birds in the sky are cormorants flying south in long strings. They too ride thermals occasionally, swirling into loops before heading out over the water. The birders' conversation has degenerated: "Can you name the movie Cary Grant and Ingrid Bergman made together?" *"Notorious."* "OK, now tell me the four movies Paul Newman made that began with an H." "Wait, first you tell me who played the Penguin in *Batman.*"

"Where's the merlin show today?" asks Vince Elia.

"Good question," says Bob Barber. "It should have started already." Merlins are an afternoon phenomenon at the Point.

"Here you go," says Bouton, a few minutes later, as a chocolate-breasted falcon pumps over the dunes and past the platform at high speed. "That's about as merlin as you can get."

Soon after, the sky darkens still more, the wind picks up, and no birds are in sight but a few gulls over the ocean. One by one the birders say good-bye to Bouton and leave. Several turn on their headlights as they drive away. In the distance, out over the Bay to the northwest, it is raining.

By 4:30, with an hour left to watch, Bouton is alone on the platform when one last visitor walks up the ramp, a bearded young man with a scope over his shoulder. A moment later, Bouton is pointing to the airspace behind the platform, then tracing a familiar pattern in the sky, first overhead from the north, then west past the park office down the beach, then around the whale jawbone to the grass field next to the parking lot. Finally, Bouton's hand wobbles up and down. The wheatear is walking again.

Chapter Five

❖

HIGBEE'S BRIT

DAWN AT HIGBEE'S BEACH, 6:45 A.M. IT HAS RAINED ALL NIGHT, AND A fine mist hangs in the air under low, heavy clouds. The Englishman Richard Crossley and CMBO's research assistant Dave Wiedner are sitting in Wiedner's car, a rusting 1978 Datsun parked in the phragmites beneath the dike. Crossley is slumped in the seat, eyes closed, raincoat zipped, the hood of his sweatshirt pulled forward; Wiedner sips coffee from a paper cup.

A man in a blue visor bends to Wiedner's window. "Is this where they're counting all those warblers?"

"This is the place."

"Are you the counter?"

"I'm the scribe," says Wiedner. "Richard here is the counter."

The man steps back from the car and lifts his binoculars. "There goes a flock of something or other. Little spots, who knows what. . . . Whoa, here come some more."

Wiedner takes a final sip and crumples his cup under his seat. "Come on, Richard. It's time."

Crossley pushes his door open and rolls out with a groan. "I could still be in me bed. I *would* still be in me bed if you had stayed away."

Wiedner leads the climb up the earthen wall to the top of the dike, the man in the visor follows, Crossley goes third. Clumps of mud cling to their shoes.

"I heard on the hot line you've seen some Connecticut warblers," says the visitor.

Wiedner nods. He is studying the flags on the ferry at the terminal half a mile to the north. He scribbles estimates of wind speed and direction on his clipboard, then removes an oversized Celsius thermometer from his backpack, tucks the backpack under a bush, and places the thermometer carefully across its top.

"*God,* I'd love to see a Connecticut," the man in the visor continues. "That's a *rare* bird."

"Not really," says Crossley. "Not here. We had thirteen in one day, twenty-five in one week."

A loose cluster of half a dozen small birds crosses overhead, murky dark against the gray sky and fifty yards up at the nearest point. "Three black-throated blues, two northern waterthrushes, one redstart," says Crossley. Wiedner turns pages on his clipboard and pencils in the numbers. A solitary bird, following the cluster but twice as high, emits a single, thin chip. "Hear that?" asks Crossley. "Add another waterthrush."

The dike is a dirt embankment, twenty feet tall, and the edge of the mud flat that was created fifty years ago when the Cape May Canal was dug and the spoil deposited here. The flat is a quarter of a mile wide and littered with tires and other flotsam that has washed up out of the Canal. Crossley and Wiedner generally keep this scene at their backs. Below and directly in front of them is a horse trail that curves gently northeastward, running away toward their left along the border of Higbee's woods to the Canal. Oaks, locusts, hollies, and hackberry trees—all of them tall, old, and thick with lichens, vines, and spiders' webs—line the trail south and east of the dike.

The dike's prettiest vista is on the right, to the west, along a line of oaks to an open view of Delaware Bay, where bottlenosed dolphins are sometimes seen and this morning two fishing boats are anchored. On sunny days the light is best in this direction. The only songbirds an ordinary birder has a reasonable chance to identify are those that fly out of these westside oaks and stay low as they cross over the phragmites immediately below the dike. If you can get your binoculars on one of these birds the instant it leaves the trees, you have three or four seconds to name the species before it flies out of range.

But the birds come out of the woods from all angles and heights, in groupings too loose and heterogeneous to be called flocks: a cluster of twenty birds high overhead; two clusters of eight at eye level; three clusters of five, all low against the gray glare in the east; a solitary bird from the base of a tree in front of the dike; two birds high overhead; a cluster of ten in the good light on the Bayside; another cluster of five in the east-side glare; a cluster of seven at the limit of vision straight up, two hundred yards high.

"Two yellowthroats, three redstarts, two parulas," says Crossley, wheeling quickly to face each group as it comes. "Two scarlet tanagers. . . . Five parulas, three yellowthroats, five spahs." "Spah" is Brit for "sp.," the scientific shorthand for "species, unidentified." "Four white-eyes, two magnolias, eight spahs. . . . Make that ten spahs. And add another maggie." Wiedner turns pages and scribbles, his own binoculars hanging unused.

Each afternoon, after completing the count, Wiedner has been entering the *New York Times* daily weather codes into CMBO's computer, trying to find some correlation between weather systems to the north and west with the count of songbirds at Higbee's that morning. "So far it seems pretty much random. We see more birds on local northwest winds. Other than that, I haven't found any pattern."

The most commonly repeated theory about Higbee's morning songbird flights is that these migrants departed at dusk the previous evening from various areas far to the north and west—Connecticut, Long Island, northern and central New Jersey, eastern Pennsylvania—and have been flying all night. Pushed off course by westward or northwestward winds, they found themselves out over the ocean at dawn, turned into the wind, and fought their way back to shore. They descend into Higbee's woods to feed on caterpillars and other insects, and then, apparently, they take off again, flying northward out of the woods, over the dikes, and up the Bayshore toward Villas, Dias Creek, and Goshen. The explanation generally given for this northbound departure is the same as the old explanation for the northbound hawk flight: "They're looking for a narrower place to cross over the Bay."

At least part of this theory is an error, however. Warblers and the songbirds that fly with them are not as reluctant to cross water as is often assumed. Few songbirds glide in flight, and none needs the land-based thermals eagles and buteos require. Warblers are actually more efficient flyers than hawks, burning much less metabolic energy per mile covered in direct flight, and they make overwater crossings of the Gulf of Maine, the Great Lakes, and the Gulf of Mexico. They are also regularly seen from deep-sea fishing boats fifty, sixty, and even a hundred miles off the Atlantic coast (where no hawks but the two seagoing falcons ever go). It's hard to believe the fourteen-mile width of Delaware Bay is such an impediment to them that they would backtrack thirty miles in search of a different route.

Witmer Stone questioned the "narrower crossing" idea in his *Bird Studies at Old Cape May.* "Such action would involve remarkable intelligence and judgment," he wrote in 1937. "Moreover, I have been unable to secure any evidence that they cross the upper Bay or the lower Delaware River at any point." In the years since, though thousands of birders have investigated both the New Jersey and Delaware sides of the Bay, no one has been able to locate any crossing point or even another concentration area north of Higbee's.

Recently, Paul Kerlinger and others have wondered if the northbound morning flight is a kind of optical illusion. "The birds coming off the ocean from the east descend as they reach the Point and spread out west," Kerlinger believes, "each bird looking for habitat appropriate to its species. The ones that find appropriate habitat are the ones we *don't* see from the dike. They find some place to feed or rest in the woods and stay there until the next night. The birds we do see from dike are the ones still spreading out and searching for feeding areas. They're heading north because that's the one direction they haven't explored yet. The reason no one has ever found a concentration point north of Higbee's is because there isn't one. My guess is they spread out along the coast, feed and rest, and then take off the next clear night, heading due south again. That's why Dave and Richard are doing two kinds of counts: first, the birds seen leaving the dikes, and second, the birds left behind at Higbee's. What percentage flies out? What percentage stays behind? We don't want dogma; we want *numbers.*"

Two short answers to the question "Why do birds migrate?" are equally valid:

Because they can.

Black-and-white warbler

Because they must.

Avian migration originally evolved, most theorists believe, because it enabled birds to expand their foraging ranges beyond their original territories, into areas where food sources were temporary rather than year-round. The ancestors of today's insect-eating migrants were tropical birds that probably first wandered nomadically and erratically back and forth short distances from the edges of their original ranges. They advanced into temperate areas when the weather was warm and insects were active and retreated when the weather turned cold and their prey disappeared. In time, evolutionary selection accentuated and fine-tuned these movements by rewarding those species and those individuals that developed characteristics— longer wings, stronger flight muscles, sharper orientation abilities— that enabled them to risk lengthier and lengthier flights away from their ancestral breeding grounds and finally adapted them to breed on the summer grounds and to bear young who could find their way back to the ancestral territory in winter.

Nesting in the north offers other advantages for tropical species besides greater food sources. Temperate-zone days are longer in summer, giving parent birds more time to forage for their young; more land mass exists in the northern half of the Western Hemisphere than in the tropics, and so competition for nest sites is less keen; and there are far fewer nest-robbers and predators outside the tropics—no monkeys or kinkajous, fewer raccoons, fewer tree-climbing snakes, fewer egg-eating birds.

Migration is obviously a successful biological strategy. More than a third of the world's species of birds and more than three-quarters of all North American species are migratory in all or parts of their ranges. A common guesstimate of the number of birds that head south on the North American continent each fall is five *billion,* and on any given clear night, spring or fall, the skies across the United States can be filled by hundreds of millions of birds on migration. One night in September 1977 a Clemson University researcher captured on radar the images of a million songbirds migrating over a single South Carolina airport. "Considering the great benefits of migrating and the ease and perfection of bird flight," the ornithologist Joel Welty once observed, "it would be one of the great mysteries of nature if birds did *not* migrate."

But perhaps the advantages of migration are a little too

obvious. The costs and dangers are generally underestimated by casual observers, and the basic pattern of migration, at least the pattern for tropical migrants, is misunderstood even by many knowledgeable birdwatchers. The man in the street's common misperception is evident in the countless cartoons and postcards suggesting migrant birds have a schedule we all should envy: half a year's work, then a quick and easy trip to six months of vacation in the sun. Migration is no quick trip. It is a twice-annual supreme test of health and adaptation, which ends regularly in death, especially for songbirds.

One widespread misunderstanding about migration among birdwatchers is evident in our sense that warblers, vireos, hummingbirds, flycatchers, orioles, and tanagers are "our" birds, native species that are driven away from their homes here by the bitter weather of winter. Long-distance tropical migrants should really not be considered "our" birds: all are descended from species that lived in equatorial regions, all have close relatives that live there year-round still, and all spend far more time in the south "wintering" than they spend up here nesting. Warblers, for example, spend twenty-five to thirty weeks in the tropics and only ten to twelve weeks on their nesting grounds. The rest of the year—four to five weeks in spring and six to eight weeks in fall—is devoted to migration. These birds are visitors to our latitudes; their homes are in the tropics. They are no more "native" species than the commuters from New Jersey and Connecticut who work in Manhattan are native New Yorkers.

Migration is also not the universal formula for success it can seem to be to temperate-zone birders, who know best the migratory species and generally know little of other tropical species. Only about half the 109 members of the warbler family, the *Parulidae,* migrate north to breed above the United States–Mexico border; the rest remain year-round in Central and South America. Most of the 319 hummingbirds that live in the New World are sedentary species; only twenty-one migrate north to nest the U.S; only four species reach Canada; only one nests in the eastern U.S. The tyrant flycatcher family, the *Tyrannidae,* is one of the most diverse and successful families of birds in the New World, with 374 different species now recognized, but seasonal migration seems not to have played a major role in that success; only thirty-five species of tyrant flycatcher migrate to the U.S. Birders who have tracked down each of the four species of tanager that nest in the U.S.—the scarlet, summer, western, and hepatic tanagers—tend to think they've "seen them all." In fact, if those four species are the only tanagers on your lifelist, you've missed 98.3 percent of the group. The four North American breeders are the oddballs in this tropical family of 236 species, most of which live in South America and do not migrate.

Orientation ability is the most obvious of the many characteristics a migratory species must develop. A juvenile scarlet tanager born in an oak that stands, let's say, behind the ticket taker's booth at Shaftsbury State Park in western Vermont must depart the area by late August or early September and fly by itself, at eight weeks of age, with no parents to guide it, down through New York, Pennsylvania, Virginia, North Carolina, South Carolina, Georgia, and Florida, then cross the Gulf of Mexico on precisely the right south-southwest track to reach the Yucatan coast and then turn southeast to continue through Mexico, Guatemala, Honduras, Nicaragua, Costa Rica, Panama, Colombia, and Brazil to arrive on wintering grounds appropriate to its species deep in the Amazon jungle, perhaps in a grove of trees on the bank of the Rio Madre de Dios in northwestern Bolivia. Come spring, it must turn around and head back. It may return to Shaftsbury State Park in Vermont, and next fall come back to the same grove of trees on the Rio Madre de Dios. And it will repeat this journey twice a year as long as it lives. One male scarlet tanager, captured and banded as an adult in Norristown,

Pennsylvania, in September 1948, was found dead in Boonton, New Jersey (less than a hundred miles away) in August 1956. He was at least nine years old when he died and must have traveled the New Jersey/Pennsylvania to the Amazon route at least eighteen times, covering more than ninety thousand miles while on migration.

Many species of tropical migrants, banding has taught us, fly a "loop" migration and so must know and follow different routes north and south. Connecticut warblers, for example, fly south from their breeding grounds in Minnesota and southern Canada to their wintering grounds in Venezuela and Brazil on the same trans-Gulf route scarlet tanagers follow. In spring, however, they take the land-based route, coming north through Mexico and Texas, then veering northeastward through the Midwest to return in June to the territories they left the previous August.

A precise sense of timing is as crucial a characteristic for migrants as orientation ability. Birds that remain year-round in the tropics can breed when local conditions are good, and many sedentary species do just that, mating and nesting after long rains, for example, when insects are peaking. Long-distance migrants must follow schedules that are determined by conditions in areas they cannot see, thousands of miles away. Tropical-based migratory songbirds begin their flights north in late March or early April, and each species has developed its own schedule, depending on the particular insects it is adapted to feed upon, the budding of trees en route whose leaves feed those insects, the distance between the bird's wintering grounds and its nesting territory, and the species' flying abilities. Timing errors are punished by death, as when the one-week-early warbler arrives on its nesting territory before the geometrid moth larvae have hatched and starves, or by loss of breeding privileges, as when the one-week-tardy warbler discovers its mate and the territory it ruled last year have been taken by a rival.

The dangers of mistiming are more subtle for the southbound migrants, but here too selective pressures are at work and different species have different schedules. Certain species are always the first to head south—Louisiana waterthrushes, worm-eating warblers, orchard orioles. (This southbound migration begins much earlier than is generally realized; Crossley has been counting migrant warblers at Higbee's since July 20.) Other species are always among the last of the tropical migrants to join the fall flight—blackpolls, bay-breasted warblers, northern orioles. The different distances each species must fly is one factor; generally the longer the distance between nesting and wintering grounds the earlier the species leaves its nest site, but there are many exceptions to this rule: scarlet tanagers flying from New Jersey to the Amazon are still here in September and October, weeks after the Kentucky warblers that winter the Caribbean have departed. The tanager may be the better flyer. Chimney swifts have a long migration route, from Canada to Peru, but they fly so fast that they can afford to linger on the nesting grounds later than many other insect-eaters. Subtle differences in flying abilities may also be the reason the timing of the southbound migrations of the barn swallow and the cliff swallow are different. These two species are very closely related, sister species in the same genus; they feed on the same kinds of prey, nest over virtually the same territory, and winter together in South America. The cliff swallow lacks the barn swallow's long, forked tail, however, and is perhaps not quite as fast or as tireless a flyer. Cliff swallows depart weeks ahead of their relatives. In Cape May they are rare after September 1. Barn swallows are still heading south past the platform in mid-October.

But orientation and timing are not all there is to migration. More than anything else, migration is a test of strength, stamina, persistence, and good fortune. The journeys of migratory birds have been compared so often to car- or plane-assisted trips that we forget a bird is not riding in a vehicle, having to worry only about following the map and sticking to the schedule. The bird *is* the vehicle, and the vehicle needs fuel and rest and can be destroyed any moment of any day by winds, fog, cold, lack of food, illness, infection, muscle failure, broken feathers, or a sharp-taloned predator. Each time the bird moves on, it encounters new puzzles to solve and new tests of its health and its spirit, and the dangers continue for weeks. How far can I fly today and still have energy left to find food? How high can I fly and not get blown off course by winds? Where can I land today and be safe from sharp-shins and merlins? Where will the caterpillars I can eat be found in this forest?

The story told more often than any other in books about migration is the story of the blackpoll warbler's southbound migration; it may be the most extraordinary traveling act on the planet. Blackpoll warblers move southeastward from their nesting grounds in Alaska, Canada, and New England to coastal staging areas from Cape Cod to Cape May. There they gather their strength, feeding on insects and building their fat reserves until they have increased their weight by a full fifty percent. Finally, one night when the sky is clear and the winds are right, they depart. They fly, pumping their wings constantly—warblers cannot glide—for eighty to a hundred hours, well offshore and high above the ocean, over the Baltimore Canyon on the edge of the continental shelf, over Diamond Shoals off Cape Hatteras, over Bermuda at a height of sixty-five hundred feet, where they meet the southeast trade winds and gain altitude, over Antigua at a height of twenty-one thousand feet, and finally over Trinidad, where they descend before landing on the coast of Venezuela, fourteen to sixteen hundred miles from their departure point. They cannot drink or feed or rest, even for a moment, anywhere along that route during their four-and-a-half-day trip. Many blackpolls drown along the way, of course, and others reach the Venezuelan coast too emaciated to find food and die on arrival, but this all-or-nothing method is successful for the blackpoll as a species, and the story of the blackpoll's migration is usually told to illustrate the toughness of this one little warbler. What the story also illustrates, however, is how tough the *other* warblers are. If the blackpoll's route is advantageous—and it must be so or it wouldn't have evolved in the first place—the other warblers' routes must be equally as difficult. What does the blackpoll gain by its annual life-or-death flight? It avoids the "normal" land-based route the other warblers fly, which must be just as murderous.

This morning the warblers are flying through the mist. "They were flying last night in that downpour we had," says the man in the visor. "I was walking my dog around Lily Lake, ten P.M. and heavy rain, and I could hear them coming over: one every minute or two: *cheep . . . cheep . . . cheep.*"

Crossley nods. "I heard them, too. They can fly through rain, though you don't hear people talking about it."

"We're going to rewrite migration theory with this warbler count," says Wiedner, "if the ink doesn't run."

"Two parulas, two northern waterthrushes, one black-throated blue, five spahs," says Crossley.

"How do you tell northern waterthrushes from Louisianas on the fly?" asks the man in the visor.

"Oh, the Louisiana has a huge bill," says Crossley. "An *enormous* bill. The whole head leans forward. . . . Four black-throated blues, two redstarts, five ruby-crowns, ten spahs."

You have to be a knowledgeable birder to appreciate what Crossley is doing. Identification of many groups of birds is easier than it looks, and any moderately experienced birder can startle novices and nonbirders with a quick I.D. of a distant duck or heron. A white, medium-sized, long-necked bird crosses the highway, and the veteran names it: "Snowy egret! See the yellow feet?" The novices in the backseat shake their heads in amazement or wonder if it is time they had an eye exam. "How do you see so much?" they often ask. Actually, the veteran has seen little; the trick is a simple two-step elimination process. The obvious field marks—"white, medium-sized, and long-necked"—limit the possibilities to just three species among all seven hundred-plus North American birds: white ibis, juvenile-plumaged little blue heron, and snowy egret. Since the snowy is by far the most numerous, the veteran's eyes go quickly to the feet for a confirming glimpse of yellow. The mental sequence from "white bird" to "snowy egret" is easily completed in a fraction of a second.

Crossley's warbler count, on the other hand, is much *harder* than it looks. A bird the size of a candy bar crosses the sky, and only exceptional birders will be able to identify even its family group before it flies out of sight. Is it a warbler—or a kinglet, bunting, sparrow, oriole, tanager, thrush, grosbeak, or vireo? Kinglets, buntings, and sparrows have slightly more rounded bellies and slightly faster wingbeats than warblers; orioles and tanagers are a little longer from bill to tip of tail; thrushes and grosbeaks are slightly heftier; vireos have very slightly broader wings and fractionally slower wingbeats. Very few birders can push the in-flight identification of a warblerlike songbird beyond these family groups, and if the bird in question *is* a warbler, the complexities have only begun. Thirty-six warbler species are regular fall migrants through Cape May. Twenty-two of these show different plumages for male and female; thirteen show a juvenile plumage distinct from both adults; and one, the yellow-rumped warbler, has half a dozen plumage variations alone. Add in the possibility of a black-throated gray, Townsend's warbler, or one of the other western vagrants flying by, and the total is more than *eighty* plumages. The differences between many of these plumages—between a female mourning warbler and female Connecticut warbler, for example, or between a juvenile blackpoll and a juvenile bay-breasted—are so subtle that the species are often misidentified by museum ornithologists holding dead specimens in their hands.

None of the features Crossley has trained himself to use compares to the snowy egret's yellow feet. The Louisiana water-thrush's "huge" bill is smaller than a pistachio, and when Crossley identifies it, it is generally moving at thirty or forty miles per hour high overhead. Crossley calls out the species, and even Cape May's most experienced birders, men and women who have studied birds all their lives, shake their heads in amazement or wonder if it is time for their next eye exam.

Crossley is watching today through new binoculars. Paul Kerlinger has paid up early on the promised stipend and turned over his pair of Zeiss 7 × 42s that Crossley has been coveting all fall. They are long, elegantly slim, roof-prism binoculars shaped like an H. He holds them rock still as he watches, his head cocked slightly to the right. "There's a swamp sparrow, I like that. . . . One ruby-crowned, one Swainson's. . . . Two black-and-whites, two parulas."

He lowers his binoculars and pulls on a pair of fingerless gloves. "Goddamn, it's cold," he says. "I've got to put me bloody mitts on."

"I've got an extra sweater in my car," Wiedner says.

"That's okay. I just like to complain. Complaining warms me up." *Up* rhymes with *soup* in Crossley's accent.

By 7:00 A.M. the air is clearing and the light improving. A van pulls in behind Wiedner's car below the dike, and six birders get out. Three minutes later, a second group, twenty-five birders in a long line, shuffle out of the woods.

"See what a monster you've created, Richard," Wiedner says. Since CMBO started listing Crossley's daily warbler counts on its telephone hot-line tape, the line has been ringing thirty times a day.

"But how many of those people know what they're looking for?" Crossley asks. "If I walked down there right now and started talking about 'primary projection,' how many of that group would know what I was talking about? One? Two? People don't want to work at birding here in the States. They don't want to have to really look at the birds."

"I don't know," says Wiedner. "It's moving that way—David Sibley, for example."

David Sibley is the young painter and tour leader who illustrated and cowrote *Hawks in Flight* with Pete Dunne and Clay Sutton. He is quiet and intense, regularly spends twelve hours a day in the field, is widely regarded as the best birder in New Jersey, and seems the only American who can stand with Crossley on the dike and identify warblers on the wing.

"David Sibley," says Crossley, "is the exception that proves the rule. He is *totally* untypical of American birders. He's more like a British birder, really. If you had fifty or sixty Sibleys over here, birding in the U.S. would be revolutionized. But you don't, and you're fifteen or twenty years behind us. You're where we were in Britain in the early seventies. . . . Of course, right now, we're fifteen or twenty years behind the Swedes."

Crossley has pushed this idea on Cape May's birders several

Higbee's Dike

times this fall: that Swedish birders are even better than the Brits. No one seems to believe him, or at least wants to believe him. The competence of Crossley and the half-dozen other Brits who have joined him here this year have been insult and injury enough to the Americans. "I don't know how they do it," Clay Sutton has said. "Every year more Brits show up here, most of them coming to the States for the first time, and all of them seem to know American birds like they've been watching them all their lives. The average British birder getting off the plane in New York on his first trip across the ocean knows more about our birds than the average American birder who grew up here."

Natural history is taken much more seriously in Great Britain than in the United States, especially by young males. The level of intensity among birders in the United Kingdom may be matched here only by urban teenagers playing basketball or by new college graduates working in their first jobs on Wall Street. It is an all-consuming passion. "I think it shows one of the real differences between our two cultures," says Sutton. "The British really study the natural world and treasure it. The birders we see here are all very visual, too. They seem to have trained their eyes in ways very few Americans ever do. All of them can draw; most of them never carry field guides; and you just don't hear about a Brit making a mistake out in the field. If an American comes walking into the Observatory saying he's seen a Swainson's warbler, I want to hear the details before I believe him. If a Brit comes walking in saying he's seen a Swainson's, I know I might as well write it down."

Crossley never carries a field guide, and he shares the general sense among the Brits that the Americans' dependency on their books is one reason for their widespread lack of competence. The American reaches for his field guide when he encounters a new bird and starts flipping pages, like a child turning to the end of the puzzle book so he won't have to think through the answer. The Britisher studies the bird. The discipline makes for a sharper eye and better memory.

One day in August Crossley encountered a pair of birders, a man and a woman among New Jersey's most experienced observers, who had their binoculars trained on a songbird that flew away as Crossley walked up.

"That," said the man, "was a golden-winged warbler."

"Was it?" Crossley asked. "What did you see on it?"

The man shrugged and turned to his friend. "Let me have your field guide."

The woman handed over her Peterson, the man flipped pages to the warblers' section, held the book at arm's length while he contemplated the golden-wing's picture for several moments, then tapped the illustration in three places. "It had this, this, and this." Then he slapped the book shut.

Crossley staggered away, he says, biting his tongue. Now he shrugs, "That's the typical American attitude. 'I don't have to remember what I noticed; I can just look it up.' The British attitude is if you have to look in some bloody book to know what you saw, you *didn't* really see it."

In single file, the field trip crowd climbs the dike's wall.

"Have you seen any Connecticut warblers today?" asks the woman leading the trip.

Wiedner shakes his head, "Not yet."

"Any Connecticuts today?" asks a man at the end of the line.

A sizzling zip—*zzzzzzzzt!*—descends from a bird so high in the sky it is not visible. "Hear that?" shouts Crossley. *"Dickcissel!"* Two dozen birders scan wildly at the sky, their binoculars wobbling in various directions. Meanwhile, the leader at the far end spots a sharp-shin over the trees and points the hawk out to the others. All binoculars swing into alignment, and the group oohs and ahhs.

Crossley shakes his head in disgust. "The reason Americans like raptors is because they're so big and slow-moving you can't miss them. There's no challenge to your ability. You can stand around here on the dike or sit on the platform all day and have a good chitchat. The whole social scene."

"Raptors can be hard," says Wiedner. "At a far enough distance."

Crossley snorts. "*Any* bird is tough at infinite distance. *Mute swan* could be tough."

Wiedner laughs.

"I'm serious," says Crossley. "You get two swans over Delaware and you wouldn't know what they were." A dozen songbirds in a loose cluster pass overhead, pumping. Crossley ignores them. "Two hundred and fifty species come through Cape May every fall. How many are looked at? Eighteen! If it isn't a hawk, no one cares about it. It's *stupidity.* It's *bloody ignorance!* Why should all the emphasis be on hawks?"

"One reason," says Wiedner, kneeling to check his thermometer, "is hawks are indicator species. The DDT years proved that. When the environment has poisons, hawks die."

"*Good!* That warms me heart. One raptor isn't worth one warbler."

"Pesticides kill warblers too, Richard."

"Why can't warblers be the indicator species, then?"

"They're harder to census. You can't see them all."

"You can't see all the bloody raptors, either!"

Ten songbirds come out of the east. "Five parulas, one magnolia, four spahs," Crossley announces. He exhales loudly. "I'm not positive about the magnolia."

"Eighty percent sure?"

"Yeah, eighty percent."

"Good enough," says Wiedner, turning over the page and writing in the numbers.

"Mind you, I take an extreme angle on this antiraptor talk. I have to—to get in certain people's heads." He looks at Wiedner. "Do *you* think they should keep the hawk banding going?"

"I think it might have to be cut back some. They should keep the demos going. That's the most important thing for PR purposes."

"This is arguably the best migration place in America, and CMBO might be the best observatory in the U.S.," Crossley says, "but it doesn't compare to any of the British observatories. All the money is going to raptors. None of the other birds are being studied. CMBO ought to do warbler counts spring and fall, wader counts spring and fall. How many waders go through Cape May each fall? Nobody knows! You don't know what's going through your own backyard! How in hell can you say something is doing well or doing badly, when you don't know what you have in the first place? You can't say 'I lost five quid' if you don't know how much money you have in the bank."

"But that's what we're working on, Richard. That's why Paul started this warbler count."

"All right, bully for him. But will he sponsor a warbler count *next* year?"

"I hope so." Wiedner points at Crossley's new binoculars. "Maybe he can give you his telescope next year."

Crossley estimates he has hitchhiked more than a hundred thousand miles since he graduated from college three years ago. He has birded in Singapore, Malaysia, Thailand, India, Nepal, Israel, Greece, Spain, France, Mexico, Florida, Virginia, California, and New Jersey. This is his fourth trip to Cape May. "Lots of British birders go to more places than I do, but they don't stay in one place for months the way

I do, and they don't come back to the same place again and again. I like to study what you can really expect in each place."

He has been fascinated with birds as long as he can remember. His father interested him in the hobby of egg-collecting, and by the time he was eight, Richard was tracking goldcrests and chiffchaffs through the woods and climbing trees to search for their nests. At age ten he won a school contest in bird identification and was invited on a field trip by the teacher who had organized the contest. The trip was something of a letdown: Crossley discovered he knew as much about birds as the teacher did. "After that I carried on on me own." A few years later, his family moved to a coastal county, Devonshire, in southwestern England, an excellent birding area, and his pursuit of birds grew still more intense. At age sixteen he told his parents he wanted to spend his October school holiday on the Scilly Isles, the cluster of islands thirty miles off the coast of Land's End that is Britain's most famous migration trap. "They were dead set against, but I said I was going anyway. That was a big step, what began my world traveling, actually."

Like Cape May, the Scillies are at the end of a funnel for migrating birds. Birds that have bred in Scotland and Wales are directed toward the Scillies by the Irish Sea on one side and the North Sea and English Channel on the other—as are birds that have nested in Ireland and Iceland, both due north of the Scillies. Off-course migrants from Scandinavia and Russia also regularly appear on the Scillies, but the islands' greatest claim to fame among British birders is its long list of New World vagrants. Westerlies are the prevailing winds on the Atlantic, and gray-cheeked thrushes, rose-breasted grosbeaks, golden plovers, buff-breasted sandpipers, and other North American birds make regular appearances on the Scillies.

On the Scillies Crossley also got his first good look at "twitchers"—British birders who travel the world in search of rarities and who come to the islands each fall to race against each other in pursuit of vagrant migrants. "That inspired me more than anything. Birding in Britain is a younger man's game, and much more competitive. You want to be competent, and you want to find the good bird first. You don't want people pointing everything out to you. That's why I came to the States, really—so I could learn North American birds well enough to recognize the vagrants back home on the Scillies."

Crossley first came to the United States in the summer of 1985, and Wiedner was one of the first Americans to meet him. "I had a job that summer guarding the piping plover nests at the Meadows. I was standing on the beach, and along comes this sunburned, five-foot-six-inch Brit, who had just hitchhiked down from Wildwood. 'Hello, mate,' he says, 'do you suppose there are any wahblahs about?' "

Crossley's pursuit of warblers led him to the dike. "It didn't take long to discover Higbee's was the best place in the Point for songbirds. It did take some time to find the best place at Higbee's was this dike. Now I don't want to be anywhere else in the morning. I'm afraid I might miss something."

He will leave Cape May at the end of September to return to England to spend the first week of October on the Scillies, the peak of the migration there and an event he hasn't missed since his first visit at age sixteen. He will winter in Japan, where he plans to find work teaching English. "Bird, work, bird, work. Build up your savings and then go elsewhere. That's the method."

By 9:30 A.M. a bright haze has replaced the mist, dragonflies are zigzagging over the phragmites, and the warbler flight has ended. Wiedner makes his last check of wind direction and temperature, then puts his thermometer away in his backpack and follows Crossley down off the dike and into the woods. "Now we do the habitat survey," he explains. "We want to know how many birds are still in the woods compared to how many we saw fly out of here: what percentage moves up the Bayshore, and what percentage stays right here in Higbee's to refuel."

Crossley is waiting for him at the first path. "Hear the Swainson's thrushes? The call sounds like water dripping."

Bik, bik.

"There," says Crossley. "Three Swainson's, two red-eyes, one white-eye, four redstarts, two black-and-whites. . . . One scarlet tanager."

Wiedner writes, Crossley marches another twenty steps and listens. By the time Wiedner catches up, Crossley is ready to report: "One black-and-white, two Swainson's, one phoebe, one solitary vireo." Wiedner turns pages, Crossley moves on.

Sometimes Wiedner must run to catch up, and he rarely has time to use his own binoculars. "That's what the scientific method is all about," he says. "The denial of personal pleasure."

Crossley makes a noise with his finger held to his mouth, like a man asking for quiet: *"Spsssh . . . sphssh . . . spssh."* The technique is called "pishing," and it attracts birds, though no one seems to know why. Chickadees fly down and surround him, calling, *dee—dee—dee.* They are nonmigrants, however, and he ignores them, watching and listening to the birds farther back in the trees: "Three veeries, one Swainson's, one maggie, two red-eyes, another solitary. Make that two Swainson's."

At one point, Crossley stops, reaches into his pocket, and takes out a piece of hard candy. "Who's going to pish them now? I'm going to have a sweet."

Wiedner pishes: *"Sppss, spsss, spssst. . . ."*

Crossley unwraps the candy deliberately, studying the twists in the paper, listening to the birds responding to Wiedner. "One veery, one Swainson's, two black-and-whites, one parula, one white-eyed," he says without lifting his head. "That's all." Then he pops the candy in his mouth and marches on.

The path swings westward through overhanging oaks out into an open area of myrtles and bayberry on the dunes overlooking the Bay. A schooner with burgundy sails is passing on the horizon. Crossley pishes, and immediately four thrashers come flapping out of the woods from different directions, their long tails pumping with the effort. One lands awkwardly in a myrtle ten feet from Crossley. It's the size of a mockingbird, with rust-red wings, a chocolate-spotted breast, and eyes the color of kumquats. *Sttittt, sttitt, sttittt,* it calls. "*Spssssh . . . spssssh . . . spsssh,*" says Crossley. Three more thrashers appear. One swoops past Crossley's knees: *Sttitt, sttitt, sttitt.* "Nothing here," says Crossley. Thrashers are short-distance migrants. "Let's head back."

They reach a meadow the size of a football field. "The sparrow field," says Crossley. "It's wet today."

"Does that mean *I'll* get wet today?"

"It means I won't. You'll have to do the honorable thing."

Holding his clipboard above his head, Wiedner wades through the waist-high brush, pishing loudly. Sparrows flush left and right, but Crossley's back is turned. The chip of a bird from high in a tree has distracted him, and he walks over to look.

Wiedner returns from his long loop with blue jeans soaked, pencil in hand. He has flushed at least fifty birds. Crossley turns around to face him and shrugs. "Put down a big round number," he says, "zero. Thought I had a Lincoln's up there. But I only heard it, to be honest. Never saw it."

On the edge of the field two hawks—a sharp-shin and a merlin—fight over a snag at the top of a dead tree. The merlin swoops down, forces the sharp-shin into the air, and takes the perch. The sharpie flies high into the sky, squawking like a parrot, then loops around, swipes at the merlin from behind, and forces it back into the air. Both species are warbler-eaters. "Go on!" Crossley shouts up to them. *"Kill* each other, you buggers!"

Back in the woods, Crossley points out a brown creeper climbing up the side of an oak. It's a tiny bird, the size of a wren, picking for insects in the bark with a bill the size of a fingernail clipping.

"Neat," says Wiedner. "I've never seen one before." He hooks his clipboard on his belt and reaches for his binoculars for only the third or fourth time all morning. "It uses its tail like a woodpecker."

"That's a new tick for you, eh?"

"Yeah."

"That's a pint of beer you owe me, then. The twitcher's tradition."

"Sure, Richard. If I believed any of that stuff, think of all the beers I'd owe you. I'd be paying your bar bill for a month."

They reach Higbee's parking lot at 11:45 and reconnoiter with half a dozen birders gathered there. "Anybody see a Connecticut today?" someone asks. Heads shake no all around the circle. One person reports a Lincoln's sparrow, and someone else an unidentifiable flycatcher: "I think alder, but it didn't call, of course." "Can you believe all these swamp sparrows?" another asks. "We must have had fifty." "More like eighty," says his companion.

Two birders trot up to the group. "Hey! surprise bird of the day—hooded warbler, spring-plumaged male. Behind the Research Station, two minutes ago."

One young man immediately breaks from the group to run up the street toward the Research Station, and everyone else follows.

When the pack turns right into the Station's front yard, however, Crossley turns left and disappears into the woods. A few minutes later, he emerges a hundred yards farther up the road, calls to the group, and points out the bird, sitting on a looping vine deep in the tangle. "I thought it might cross the road," he explains. "A lot of flocks circle around here that way."

It's noon. Wiedner must return to the CMBO office to enter the morning's data into the computer. Crossley has hitched a ride to Freda's Lunch Shop with one of the other birders.

"That was a good morning," he tells Wiedner with a wave. "Thanks for getting me up."

"My pleasure, Richard," says Wiedner. "See you tomorrow—same time, same place."

Chapter Six

❖

TALKING
NUMBERS

ORDINARILY, A REGULAR SEQUENCE OF EVENTS MARKS SEPTEMBER'S TURN to October at Cape May Point: the children and the sunbathers desert the beach; phragmites flower heads turn purple; laughing gull heads turn gray; shorebirds and warblers depart for points south; sparrows and scoters, the oceangoing ducks of winter, arrive. Then, one evening, storm clouds gather, windows are closed, doors are locked, oil burners rumble to life, and a cold, cold rain drums all night. The next morning, all watchers on the platform are wearing gloves and ski caps, puffy white clouds are racing toward Delaware, and the hawks are pouring by.

Not this year. This year the last week of September was as sunny and front-less as the rest of the month. Other birds, less weather-dependent than hawks, moved as usual. The shorebirds have deserted the Meadows—their molted feathers hang in the reeds there now like litter in Sunday morning streets after a Saturday parade. The warblers have also come and gone. Sparrows are everywhere around the Point—every roadside hedge seems to hold at least half a dozen. And the first flocks of scoters, loose strings of black beads, have appeared on the horizon. But the hawk flights have been mediocre.

Each morning since mid-September a few birders have climbed the steps of the platform in gloves and ski caps, as if they could will the weather to change. By ten or eleven—about the time the first sunbather sashayed over the dunes in a bikini—they were sitting back on the bench, eyes glazed, hope gone, their gloves and caps stuffed under their seats. "Maybe tomorrow."

"Maybe tomorrow" has become this fall's mantra. Any time Bouton's weather radio reports a front passing through New England or across Lake Michigan, all listeners on the platform sit up. "Hey, sounds good." "Yeah, maybe tomorrow." "God, we need some weather." "We need a *big* flight." "Yeah, maybe tomorrow."

Shorebirds and warblers, which have farther to travel than hawks, cannot linger. Sparrows migrate at night, without the aid of thermals. Scoters and other ducks fly in any weather. But hawks need cold fronts, and hawkwatchers in an Indian Summer are as cranky as gamblers in Bust City.

"Where the hell are the goddamned hawks?"

"Maybe tomorrow."

"Fuck that. I have to be at work tomorrow."

A couple of late September events felt like bad omens. The first Swainson's hawk of the season arrived so emaciated that the bander who trapped it fed it her lure bird, something banders never do. The hawk consumed the pigeon, but then lingered in the area for three days, flapping weakly here and there around the Point, so obviously a doomed migrant that watching it was depressing. Worse was the screech owl discovered one morning standing at an opening in a hollow oak at Higbee's Beach. All day long it stayed at the opening, while one birder after another returned to the platform to

"Tomorrow is here."

report its presence: "So red . . . so tame . . . so cute. . . ." Late in the afternoon, the twentieth or thirtieth person to go look at the owl grew suspicious. He walked across the field until he was directly beneath the bird, then he reached up and yanked it down. It was dead, and had been for several days. It was a road-killed bird, apparently, propped up and tied to the tree with fishing line by a prankster.

Today, October 5, we finally have hawks. The first moderately strong front of the season passed through yesterday afternoon and last night, and this morning the platform is packed with expectant birders— "Tomorrow is here!" Out over the ocean ospreys are pumping by in ones and twos; above them harriers are gliding out, their wings and tails in crossbow silhouettes. Sharp-shins circle directly over the platform, then beat their short wings as they reach the water's edge, each fluttering as if desperate, like a man waving his arms backward to keep from falling off a cliff, and they climb higher and higher, gaining altitude for the crossing. Inland from the platform broad-winged hawks are riding in kettles of five and ten birds. Each kettle follows the same path—floating westward out from the Meadows, then veering west-northwestward over the woods and the parking lot, to pass inside the Lighthouse. As they continue down the peninsula toward Sunset Beach, not a single broad-wing enters the airspace over the water. Broad-wings are buteos and need stronger thermal winds to make the crossing. Today they will turn north when they reach the Bay and head back against the path they have followed to reach the Point.

"Where's Sutton when I need him?" Jeff Bouton calls out. The larger accipiter circling with the sharpies—with banded tail; barred breast; gray, two-toned wings—is either a big female Cooper's or the first goshawk of the season. It rides in a blue and wispy-white sky, higher than some cloud fragments below.

"It's a gos!" shouts a man from the left-hand corner of the platform. "Look at that big chest. A *huge* chest."

"Too much white in the belly," answers a man on Bouton's right. "A gos should be streaky all the way down."

"Look how short the tail is."

"You call that short?"

The hawk spirals upward and outward—flap, flap, glide; flap, flap, glide—until it crosses directly through the sun, and the watchers must drop their binoculars. The goshawk advocate goes to his backpack, returns with his field guide, and reviews the field marks with Bouton, tapping his finger on the page. "It had this . . . this . . . this." The Cooper's man looks on silently, then walks to the far end of the platform, and scans the ducks in Bunker Pond.

"OK," says Bouton. "It's official. We're calling it a gos. I just want to be careful. Gos is the one bird that gives me trouble. I don't see enough of them."

"I've seen a million of them at Duluth," says the advocate.

Sutton arrives a few minutes later, too late to see the goshawk, and noting the crowd, he does not walk up onto the platform. Instead, he leans against the bike rack next to the sidewalk and reads the Totals Board through his binoculars.

Despite the hawks overhead, Sutton is glum. "If we're going to have a good count this year, something has to happen soon. It might already be too late to save the count this year. We've just gone through what should have been the best ten days of the season without a single good front. The osprey count is OK, merlins aren't too bad, harrier and Cooper's are respectable. And that's it. Everything else looks down. The fronts we get in October aren't going to bring us the birds we missed in September. Those birds have moved south already. It's not the total of fronts you get each fall that matters so much; it's *when* they come.

Talking Numbers

❖

Cape May Point Hawkwatch

SPECIES	YESTERDAY'S TOTAL	TOTAL TO DATE	PEAK FLIGHT AND DATE
TURKEY VULTURE	4	76	17 9/30
GOSHAWK	—	—	—
COOPER'S HAWK	22	503	117 9/28
SHARP–SHINNED HAWK	142	4,401	825 9/28
RED–TAILED HAWK	—	83	16 9/15
RED–SHOULDERED HAWK	—	4	2 9/28
BROAD–WINGED HAWK	4	725	134 9/22
ROUGH–LEGGED HAWK	—	—	—
SWAINSON'S HAWK	—	1	9/26
GOLDEN EAGLE	—	—	—
BALD EAGLE	—	14	3 9/22
NORTHERN HARRIER	41	576	49 9/24
OSPREY	152	2,144	283 9/26
PEREGRINE FALCON	12	225	51 9/28
MERLIN	66	1,167	190 9/26
AMERICAN KESTREL	122	4,970	833 9/21
BLACK VULTURE	1	1	9/21
TOTAL	553	14,890	

"Sharpie is the species that worries me the most. That's our bread-and-butter bird. Over the first ten years of the count we averaged thirty-eight thousand a season; we had sixty thousand in nineteen eighty-four. There's been a steady drop every year since. Last year we couldn't break twenty thousand, and unless something radical happens in the next week or two, we're going to be somewhere down near twenty thousand this year. Maybe even less. Sharpie was down on just about every count along the coast last fall—Shinnecock, Fire Island, Kiptopeke—and that's been the trend the last three or four years. I don't think you can say it's only the weather. I think something may be wrong with the sharpie population."

Not all biologists believe hawkwatch data can be trusted. Some believe that extrapolating population trends from year-to-year changes in lookout counts is a guessing game for self-deluded amateurs. Weather is only the most obvious variable. Another is "observer bias." Put a better birder on any watch, argue the cynics, and the populations for all species will go up. Bring in someone less experienced, less determined, or less attentive, and the numbers go down. All but the top half-dozen sites in the country rely on unpaid volunteers, and the attention to the details of data gathering—estimating wind speed, recording sky codes, counting individual hawks in each kettle, adding up the hourly totals—varies widely from one counter to another, and more so from site to site. Some sites run their censuses dawn to dusk August through December; others count only during midday for a few weeks in September and October; some lookouts are manned only on weekends. Most sites have short histories, ten years or less, and change their schedules and procedures annually; others operate intermittently, closing down when the local organizer loses interest or moves away. Coverage is also heavily biased toward the Northeast, especially New England, and extremely spotty west of the Mississippi. New Hampshire had twenty-two sites reporting their fall 1987 totals to *Hawk Migration Studies,* the journal of the Hawk Migration Association of North America and the clearinghouse for all reports. Not a single census was conducted in Montana, Idaho, North Dakota, South Dakota, or Wyoming that fall, although there are probably more nesting hawks in each of those states than in any northeastern state. Only one season-long count was conducted in all of California in the fall of 1987, at the Golden Gate Bridge, and only one in Nevada, at the Goshutes on the western edge of the Great Salt Lake Desert, yet each recorded eleven thousand raptors, twenty-five hundred more than the total for all the New Hampshire sites combined.

For these reasons and others, professional biologists generally disregard migration-count data, no matter what sites are supplying the numbers, and ornithological journals and symposia rarely accept a paper based on the counts alone. Articles about raptor populations in refereed journals are most often based on counts of nesting pairs in known breeding territories or on the data from the national Breeding Bird Survey organized by the United States Fish and Wildlife Service each summer.

Hawk counters, for their part, show no more respect for academics' birding abilities than the academics show for the counters' data. At most lookouts, calling someone an "ornithologist" or an "academic-type" is a way to suggest that he may know Latin terminology but out in the field he can't tell a falcon from a frisbee. Hawkwatchers believe ornithologists do too much of their birding from armchairs. One day on the platform, a birder back from a trip to the South told Sutton that a pair of Mississippi kites had been discovered nesting in a suburban neighborhood of a Florida city, a block away from the home of a nationally known ornithologist. "Did *he* find them?" Sutton asked. "No," said the birder. Sutton nodded, "I rest my case."

In defense of their numbers, hawk counters argue that

weather changes and other variables should cancel each other out over the long run, and to prove their charts can measure real trends, they point to their data for three species: bald eagle, osprey, and peregrine falcon. These three raptors suffered huge losses soon after DDT was introduced in 1946, dwindled in numbers to all-time population lows in the late '60s and early '70s, and have been recovering since DDT was banned in 1972. The counts from Hawk Mountain, the only hawkwatch in the country in operation through the entire DDT period, reflect the decline of all three species, and most eastern hawk-watches in operation over the last fifteen or twenty years have data that show a steady increase in their numbers. Cape May's average counts for bald eagle and ospreys have doubled since the mid-'70s, and the peregrine count has quintupled.

A striking correlation of migration data to population trends comes from Hawk Mountain, where observers have been keeping separate totals for immature bald eagles and adult bald eagles since the watch began in 1934. The white head of the adult eagle is an ideal field mark—visible nearly as far away as the bird can be identified and present only in birds four years old or older—and so the crash and recovery of the species is dramatically reflected in the half-century of counts there. The numbers of both age groups held steady through the 1930s and '40s before DDT. The first sign of trouble came in the early 1950s as the totals for immature eagles declined, reflecting (we know now) nesting deaths due to eggshell thinning, addled eggs, and other pesticide-induced reproductive failures. The counts for adults declined next, in the mid-1950s, as a diminishing number of young birds reached maturity and the population failed to replace the adults dying from long-term poisoning and natural causes. After DDT was banned, the numbers of immature eagles went up first, in the early 1970s, as more and more nestlings fledged. The counts for adults didn't go up until the late '70s, when the new, post-DDT

cohort of young eagles finally reached adulthood. All five trends—
both age classes steady, immatures down, adults down, immatures up,
adults up—are evident in the Hawk Mountain data tables.

"What we're really doing," hawk counters like to say, "is
measuring the quality of the environment. When something is wrong
out there, we'll know about it first. Hawks are ecological litmus
paper. The DDT years proved that."

Bald eagle, osprey, and peregrine falcon share other characteristics besides their susceptibility to organochlorine pesticides, however. They are also three of the least water-fearing and least
weather-driven of all North American raptors. Each follows a restricted and relatively predictable migration route, each is a strong
flyer, and none of them migrates in flocks. A *steady trickle* describes
the migration of each, and so a few days of good weather or a few
days of bad weather won't alter a season-long count very much.
Ospreys rarely reach three-digit totals on a day's flight anywhere but
Cape May, peregrines and bald eagles rarely reach two-digit totals
anywhere else. All three are also easily identified and among the most
charismatic of all birds, species that even inexperienced and unalert
watchers count with care.

Other raptors are much harder to census. The most prob-
lematic species in the East is the broad-winged hawk, which is both
the most numerous of eastern hawks and the least predictable. At
inland lookouts it generally outnumbers all other hawks combined
and regularly occurs in single-day flights of two or three thousand
birds. But its flight path changes as often as the wind, and all sites'
season-long counts vary enormously from year to year. Even die-
hard hawk counters concede broad-wing totals should not be tied to
population trends. Hawk Mountain has counted fewer than four
thousand broad-wings some years and more than eighteen thousand
in others; the Cape May seasonal total has varied from one thousand

to thirteen thousand; the Montclair watch has seen fewer than fifteen hundred in one season and more than forty thousand in another. In 1983, Montclair and Hawk Mountain each counted fewer than seven thousand broad-wings for the year, mediocre totals for both sites, and Cape May had less than nine hundred, its lowest total ever. It was no down year for broad-wings, though. On September 14, 1983, observers at Scott's Mountain, a little-known and irregularly manned lookout in northwest New Jersey, counted 18,500 broad-wings in a single watch—as many broad-wings in one day as Hawk Mountain had ever had in an entire fall in any of its fifty seasons. September 14, 1983, also happened to be the first day of operation of a new fall watch at Morgan's Hill, Pennsylvania, southwest and down ridge from Scott's Mountain. By afternoon's end, 10,063 broad-wings had been counted there. Were some or all the Morgan's Hill birds the same as seen from Scott's—or were there twenty-eight thousand broad-winged hawks on the Kittatinny Ridge that day? No one knows. All that is certain is that most broad-winged hawks swept south that fall without being seen by any of the most carefully monitored sites. Most broad-wings probably escape the counters every fall. Both Morgan's Hill and Scott's Mountain are still irregularly manned, scores of other lookouts remain entirely unmonitored, and in nearly every issue of *Hawk Migration Studies* regional editors beg their readers to stay away from the big-name sites and go count somewhere no one else is watching. "It's a problem we can't seem to solve," says Sutton. "Hawkwatchers want to talk with other hawkwatchers, and they want to be where they have the best chances for a big flight. So, we have ninety-five percent of all available counters watching from five percent of all available sites."

Out West the most numerous and least predictable of migratory raptors is Swainson's hawk. Like the broad-wing, Swainson's is a water-fearing, weather-driven buteo, dependent on thermals to carry it south, and so it follows no regular route. Swainson's hawks build into kettles larger than those of any other North American hawk—single-day counts of fifty thousand birds, "smoke clouds of raptors," have been reported in Texas; they also have the longest migration of any hawk on earth, from the species' northernmost nesting grounds in central Alaska and the Yukon Territory to its wintering grounds on the Argentine pampas. Monitoring the Swainson's passage through the United States is even more difficult than monitoring the broad-wing's passage, however, because the plains and prairies of the western states generally lack the "leading lines"—mountain ridges, coastlines, and lake shores—that direct and focus hawk migration in the East. Some years no large flocks are found. In 1987 the total number of Swainson's counted by hawkwatchers in the U.S. was sixteen hundred birds, almost all by a single observer who happened upon several flocks feeding on grasshoppers in some open fields in Colorado. Yet in Panama City, Panama, where the Central American isthmus funnels their flight, 350,000 Swainson's hawks can be counted in a single season, and the total North American population may be more than a million.

Like broad-wings, Swainson's hawks have flashy underwings—bright white edged with a thin line of darkness—a pattern curiously similar to the underwings of the white pelican, whooping crane, wood stork, and several other soaring migrants. The pattern is perceptible at long range, especially evident when a bird is directly overhead, and may serve as a signaling device. *Come fly with me,* it seems to say. "Remember," notes Paul Kerlinger, "thermals are invisible. If you're a buteo starting a day of migration, you've got two ways to find out whether the updrafts are good and it's time to go. You can flap around until you find a thermal by chance, or you can watch for birds already in the air, soaring. Once you're up there yourself, you want other birds to join you because then you can judge the strength of other thermals in the sky—by watching how the other birds riding those thermals are doing. Which birds are climbing slower than you? Which birds are climbing faster? Is that bird over there climbing so much faster that it would be worth the energy it would cost to flap over and get on that same thermal he's riding?"

Kettles of broad-wings, Swainson's, and other soaring hawks tend to build as the morning goes on, as smaller kettles meet and join forces, and tend to break up in mid-afternoon, as ground and air temperatures move toward equilibrium and thermals weaken. Because every individual in the group is making individual decisions about the direction of its flight, a kettle of hawks has a different dynamic from other flocks of birds. Geese, duck, crane, and swan flocks operate by collective agreement, and are long-term units, often containing family bands of parents and young. All individuals in the flock roost together at night and depart together the next morning, often calling and circling while the group assembles, and the same individuals may fly together for weeks. Most species of hawks cluster only on the wing and only temporarily. They are solo travelers and solo hunters and maintain their distances from each other at night and depart individually the next morning. No hawk migrates with its young or even with its mate. The fidelity of those migratory hawks said to be "mated for life" is actually fidelity to a nest site. At the end of the summer, after the young have fledged, male and female hawk fly south independently and winter alone. Their reunion the next spring is evidence of their orientation ability and sense of timing, not their affection for each other.

The ephemeral nature of hawk kettles amplifies the erratic components of their flights, supports the critics' contention that migration counts are too prone to chance variations to be valuable, and frustrates all hawk counters who want their hard-earned numbers to be taken seriously. Finding ways to count more hawks and standardize annual totals is a regular subject of debate in *Hawk Migration Studies.* One correspondent suggested that Hawk Mountain's management could attempt to solve the problem of the missing broadwings by a redistribution of manpower. Rather than allowing the usual crowd of about two hundred people to gather on North Lookout, he proposed the sanctuary allow only one person to stand there. All other observers would be spaced one mile apart in a two-hundred-mile line running east and west from that point. Similar lines of hawkwatchers, so-called "close site studies," have been proposed for the south shore of Lake Ontario, the South Jersey peninsula, and the Florida panhandle. Even if such studies were feasible (a cross-Florida watch would require seven hundred observers spaced one mile apart from Jacksonville to Pensacola, all of them willing to stare into the sky eight to ten hours a day for three or four months) they could not be comprehensive unless each participant was supplied with radar equipment to count the birds flying too high to be seen from the ground. A comprehensive *national* count is beyond imagining. An editor added all totals for all species from the 126 lookout counts that reported to *Hawk Migration Studies* in 1987 and came up with 824,517 hawks counted nationwide. That may sound like a lot, he noted, but it's a paltry total when you realize observers in Panama City can count two and a half million hawks from a single site, and uncounted tens of millions more North American hawks end their migration well north of the Central American isthmus.

All this fussing would seem silly if other raptor censuses were more effective. Despite the logistical flaws and their lack of credibility in academia, however, hawk-migration counts are probably better measures of the general health of hawk populations in North America than either of the more widely respected census techniques, the *nest-site study* and the *roadside survey.*

In a nest-site study, observers return again and again to small, carefully restricted localities to count the paired hawks in the area and then monitor them through the weeks of brooding and feeding to see how many young fledge at each nest. Researchers usually focus on a single species, limit their investigation to relatively small areas (a single wildlife refuge or a single county), and often band the adults and the fledglings so individuals can be distinguished and tracked. Unless an adult is observed in breeding activity—copulating, building a nest, sitting on eggs, or feeding young—it is considered only a "probable" or "potential" breeder, and no nest is considered successful until the fledglings fly. Certainty is won with this technique; breadth and efficiency are lost. A typical nest-site study in Pennsylvania found nineteen Cooper's nests in three years. Another in Connecticut found seven Cooper's nests in three years. Using a modified form of the nest-site study, census-takers for a breeding bird atlas project in Pennsylvania located only ten pairs of northern harriers in the entire state over the five-year length of the project, although the state's harrier population was estimated to be between two hundred and four hundred pairs. Nest-site studies are best used to learn about a hawk's breeding biology—nesting behavior, seasonal timing, the influence of predators, and so on. As broad indicators of the health of raptor populations, they have limited value.

The second method, the roadside survey, is a looser, faster kind of census, and is the system used by the United States Fish and Wildlife Service on their Breeding Bird Survey conducted each summer by more than two thousand observers across the country. Each participant is assigned a route with fifty stops, one every half-mile, to be followed six times during the nesting season. The observer stands by his car, looking and listening, for three minutes at each station, counting all birds seen or heard, and then moves on. No attempt is made to determine whether a bird flying overhead has a nest in the area, and birds identified by call or song are not distinguished from birds identified by sight. Each BBS participant covers much more territory than the typical nest-site surveyor, and the BBS has become the standard method for measuring songbird populations in North America. Hawks don't sing, however, and they're warier of humans near their nest than songbirds tend to be, and so roadside census-takers generally have little success counting hawks in numbers. The average statewide total of red-tailed hawks found in New York each year by the approximately one hundred BBS surveys conducted there annually is about sixty birds. The actual red-tail population in the state is probably more than six thousand birds.

The different kinds of validity of these different kinds of censuses—nest-site study, roadside survey, and migration count—are like those that might be found by three teams of observers using equivalent methods to study, let's say, the popularity of various brands of pickup trucks in the northeastern United States. A team of "nest-site" truck researchers would restrict their investigation to one selected neighborhood and make a diligent, door-by-door count of every garage in the study area. They'd return to the same neighborhood at least weekly throughout the study period to determine whether any new trucks have been purchased or old trucks sold. The second team, the "roadside surveyors," would make no attempt to census all garages in any one neighborhood. Instead, they'd drive down one wide avenue and stop every couple of blocks to count all vehicles they could see—pickup trucks, motorcycles, station wagons, buses, moving vans. Following the protocol of BBS census-takers, they would ride their route survey six times in a year. The third team, the "migration counters," would use the simplest technique of all. They'd set up a daily, three-month watch at a lookout at a major funneling point, at the tollbooths at the George Washington Bridge perhaps, and count all pickups that pass by.

At a glance, the first team's method might seem the most

scientifically valid and the third team's the least. Both Team Number 1 and Team Number 2 have more carefully limited the range of their inquiries—all pickup trucks in a single neighborhood and all vehicles found along a single avenue, respectively—and so both teams can make more definitive statements about the populations they have studied than can Team Number 3. Team 1's garage-by-garage system would enable these researchers to be very certain that any changes in numbers they detect reflect actual changes of the truck population in their study area; Team 2 would be somewhat certain. Team 3, the tollbooth team, cannot precisely identify what geographical area they are surveying, and the data they collect seem the most susceptible to uncontrollable variables and outright errors. Like hawk-migration counters, the tollbooth team would have to contend with the problems of duplicate counting (the same individual truck being counted on successive days as two or more different trucks) and of observer bias (observers with better eyes would find more trucks than those with poorer vision; less motivated observers would likely grow tired or bored by the traffic and so underestimate the totals; and so on). Most important, the tollbooth team's methodology, like migration counters' methodology, ensures that their census must be a rather random sample. The trucks crossing the bridge are coming from an undetermined number of different localities and heading toward unknown destinations. And, if we overlook the fact that, unlike hawks, trucks have license plates, we can imagine the tollbooth watchers unable to answer questions such as "How many of these trucks come from local towns and pass by daily?" "How many are one-time crossers coming from long distance?" "How many trucks are crossing the Hudson on other bridges?" and "What percentage of pickup trucks in the northeastern United States never cross the Hudson?"

But, if the ultimate purpose of all three studies is to measure the popularity of pickup trucks throughout the northeastern states, Team 3's data are probably the most useful. Teams 1 and 2 can produce precise and apparently accurate results. Team 1 might conclude with certainty, for example, that Brand X pickups are 2.8 times as popular as Brand Y in its study area; Team 2 might find that Brand Z's popularity has declined by 6.3 percent from one year to the next along the route surveyed. The question neither team can answer with confidence is the same question nest-site researchers and roadside surveyors cannot answer: "Are your results true for the population as a whole?" The smaller and more localized the study, the more it is susceptible to small and localized effects. Brand X pickups might be more popular in Team 1's study area because it's a poor neighborhood and Brand X is an inexpensive truck—just as kestrels might be twice as common as sharp-shins in a certain study area because it's in an agricultural county with many open fields and few deep woods. Brand Z might decline in popularity along Team 2's survey route because a local dealer has gone out of business—just as a local osprey population can decline if the largest lake in the area becomes polluted. Ideally, a site for a small-scale study should be selected because it is representative of the whole, but that is generally impractical. Finding a representative study area can require a preliminary investigation longer than the study itself, and once the study begins, the researcher must find a way to control for all local variables. Generally, study areas are selected for other reasons. Both nest-site studies and roadside surveys are most often conducted by researchers who have easy access to them. Areas that are difficult to reach, whether representative or not, are seldom surveyed.

The fact that a lookout count is a random sample actually *enhances* its value as a measure of population trends. Because the lookout is located at a funneling point, we can be confident that the hawks counted are coming from a mix of habitats and a wide variety of localities, and we can be more hopeful that localized effects will cancel each other out. The sheer size of the sample censused by a lookout count—hundreds or thousands of times as large as the samples from nest-site or roadside surveys—is a further check against aberrations. The data from individual small-site studies can be combined to increase the sample size, of course, but so can the data from migration watches. A researcher trying to measure the breeding success of sharp-shinned hawks throughout the northeastern United States can attempt to find representative sites in a dozen different states, train and organize study teams to be sent to each site, and hope they find fifty or sixty pairs to track over the two-month nesting season; or he can spend a couple of hours one afternoon collecting the records of the forty thousand sharp-shins that have been censused annually for each of the last ten years by the watchers at Cape May, Hawk Mountain, and Kiptopeke.

One other large advantage of lookout counts may be their most often overlooked scientific value: lookout counts get repeated much more frequently than small-site studies. Small-site raptor studies are generally academic projects—conducted by researchers under publication pressure or a grant deadline to draw conclusions from the data gathered and write up the results as soon as possible. In academia, publication is primary; fieldwork only the requisite drudgery. The senior member on a group project sometimes doesn't even visit the site, the fieldworkers are often graduate or undergraduate students with limited experience as hawkwatchers, and the length of the study is generally less than a year and rarely more than two years. The value of any project is measured by the reception the written report eventually receives: "Was it published?" "What is the reputation of the journal that published it?" and "How often has the article been cited in other publications?" Academic researchers are also under pressure to strike out in new directions—to investigate questions that haven't been investigated before or to ask old questions in new places. Field-work that does not lead to a new publication is considered wasted effort, and since it's generally harder to publish a second article based on a repetition of the same fieldwork in the same place, study sites are often abandoned once the original data have been gathered. For all these reasons and others, small-site raptor studies generally have very short histories.

The priorities are just the reverse on a lookout count. For hawkwatchers the fieldwork is the primary motivator; the report the necessary drudgery. The audience an academic most wants to impress are the editors of prestigious journals; the audience a hawkwatcher most wants to impress are the other birders on the platform. Frank Nicoletti, the most dedicated hawkwatcher Cape May has ever had, could hardly bear to write up his reports. Even the hourly compilation was a struggle for him. Every moment with head bent over his clipboard was a moment wasted—a moment when a peregrine might sneak by. Jeff Bouton tabulates less reluctantly, but thanks to Nicoletti's legacy and to the reputation of hawkwatchers in general, Bouton's contract stipulates that he receives only half of his $1,500 pay in regular increments; he won't be paid the second $750 until after he submits his final report. The arrangement is necessary from CMBO's point of view because hawkwatchers generally feel little motivation to write up their results. The only ambition hawkwatch-ers can be relied upon to demonstrate consistently is the ambition to count for another day. This may seem like a shallow motive, but it generates long-term observations seldom matched by academic fieldworkers. Where academic researchers take pride in the number of different studies they have described and how often their articles are cited by other researchers, hawkwatchers take pride in how often they have done the same thing again and again—how many years and how many hours they have served at their watch and how many raptors they have counted. Gerry Smith, who has been studying the spring hawk migration at Derby Hill, New York, on the eastern shore of Lake Ontario since 1967 and has kept daily counts there each spring since 1979, has earned his reputation as the Lou Gehrig of hawk-watching. Smith can speak modestly about this record: "All it proves is I'm the most consistent idiot currently keeping counts," but he also defines himself by it: "I am the Derby Hill Bird Observatory."

Migration sites also earn their reputations by the length and size of their counts. The question of whether Hawk Mountain or Cape May has better claim to the title of North America's premier autumn hawk count is determined in most birders' minds by whether they believe length of service or the number of hawks seen should be the principal criterion. Only an academic would suggest measuring Hawk Mountain's reputation against Cape May's by noting how often each lookout's census has been cited in refereed journals.

An example of how reputations are made or broken and how questions are addressed in hawkwatching circles comes from the rivalry which has developed in recent years between Gerry Smith at his Derby Hill lookout and Frank Nicoletti at his new watch at Braddock Bay, also on the Lake Ontario shore, sixty miles west. The apparently contradictory totals Smith and Nicoletti have been report-ing in recent years have led to a debate about each observer's abilities and reliability, and about the nature of Ontario's hawk migration. It is no trivial matter. Lake Ontario's hawk flight is the largest spring raptor migration in the East, involving at least sixty to seventy thousand hawks each year—most of them adults heading north to nest in Canada or northern New York.

Lake Ontario is an impediment to avian migration because it is so cold and so large—seventy-five hundred square miles in area, approximately the size of the state of New Jersey—and because its longest line runs almost directly east–west. Northbound birds must attempt a direct overwater crossing of thirty-five to forty miles or turn at the shore and follow a path perpendicular to their intended direction. Until Nicoletti arrived on the scene, the theory about the hawk migration around the lake was a simple one and seemed to be confirmed by the numbers and behavior of the raptors seen at the three lookout counts on the lake's southern shore: Smith's lookout at Derby Hill in the southeastern corner of the lake, the Braddock Bay lookout on the south-central shore, and the Grimsby, Ontario, lookout in the southwestern corner. The hawks seen at Grimsby are generally flying westward, apparently circling the lake clockwise; those seen at Braddock and Derby Hill are invariably heading eastward, circling counterclockwise; and very few hawks are observed heading out over the water at any of the three sites. The lake's waters are too cold to allow thermals to form, and so it seemed that northbound hawks simply turned east or west depending on the prevailing winds of the day, then used the shoreline as a directional aid as they worked their way around to the north again. For the six seasons before Nicoletti took over at Braddock, Grimsby averaged 15,000 hawks a season; Braddock averaged 40,000; and Derby Hill averaged 50,000. These ratios were exactly as expected. The lower count at Grimsby was explained by the fact that the prevailing winds on Ontario's southern shore are westerlies. It also seemed to make sense that Derby's count was higher than Braddock's since an eastbound flight should build as it is funneled eastward. Those hawks coming north through the Finger Lakes region of central New York east of Braddock would turn right as they encountered Lake Ontario and join the eastbound flight—and so be missed by the counters at Braddock but be seen by Smith at Derby Hill.

But then Nicoletti took over the watch at Braddock Bay. In 1986, his first year, he recorded 69,000 hawks, more than double the 32,000 Smith recorded that year. In 1987, Smith had another mediocre year, reporting 38,000 hawks. Meanwhile, Nicoletti reported *106,000*—the highest total ever recorded at a single-observer hawk-watch north of Texas. Virtually all the birds Nicoletti sees are heading due east along the shoreline, but Smith has not come close to equaling the Braddock count in any season since Nicoletti arrived. Is Smith losing his edge, after his long string of watches, and missing birds he should be seeing? Or does Smith's greater experience suggest that he is the more reliable observer and that Nicoletti's totals are careless overestimations? Or could it be that both observers' totals are accurate and the old theory about the Ontario flight is wrong? Perhaps the hawks are doing something unexpected between Braddock and Derby—cutting back south, against the grain of their northbound movement, and heading inland, or shortcutting north, defying the standard assumptions of migration science, and crossing over the lake's cold water. "We might be able to settle those questions," says Clay Sutton, "if we could convince Gerry and Frank to switch places with each other some year—and then compared their totals. But that's not going to happen. Both of them are too loyal to their own sites, and caught up in the tradition of putting in another year at the same place. All we can do is see how their counts go over the next five or ten years."

What separates the typical migration counter from the typical researcher are a passionate loyalty to a single site, the time to take the long view, and a fierce sense of competitiveness. "I've been watching raptors at the Point each fall for the last fifteen years," says Clay Sutton, "and I have never even spent a single day at Hawk Mountain. That's how I know Frank and Gerry wouldn't switch places." The Braddock/Derby puzzle won't be settled until one or the other counter has convinced enough visitors to his site that he is the more reliable observer—or until both have proven their credibility is beyond question.

So it goes at lookouts everywhere, and the data base grows each year. It is already enormous, far larger than any university-based or government-based raptor study. Open any issue of *Hawk Migration Studies* and you will find more records of far more hours of observation than you will find in two or three issues of any professionally published journal on raptors. Here on a single page is the data from the most recent counts from Braddock Bay and Derby Hill: different totals for each of fifteen species of hawks for each of the last eight seasons—more than 600,000 individual sightings in all, collected during more than 10,000 hours of observation. Turn another page and you find the past seven years of counts at Grimsby, 4,000 hours of data collecting, 100,000 individual observations; here on a third page are the last seven years of counts at Whitefish Bay, Michigan, 3,000 hours, 125,000 sightings. "And we're so damned cheap," Gerry Smith has pointed out. "A few of us get paid a pittance; most hawkwatchers do it for nothing; and lots of us put in hundreds of hours a year, year after year. Hawk-migration counts are the most cost-effective environmental early-warning system in existence."

A few professional ornithologists have begun to look past the standard criticisms of hawk-migration counts and to argue that lookout data are an untapped and unacknowledged resource. "People are always pointing out how migration counts are influenced by observer bias, weather variations, lack of standardized procedures, and the other things that make them chancy," says James Mosher, a veteran nest-site surveyor and nationally recognized expert on accipiters. "But the other most common censuses, roadside surveys and nest-site surveys, drastically *undercount* raptors. I call it a visibility bias. The fact is it's so hard to find hawks on their nesting grounds that we don't know much more today about their distribution and their breeding densities than we knew forty years ago." Mosher has designed a surveying technique using tape-recorded playbacks of owl calls to lure hawks out from their hiding places, and the system has been adopted by other teams of researchers throughout the Northeast. "The tapes have improved detection rates significantly, but we're still missing a lot more hawks than we're finding. I estimate our detection rate for Cooper's hawks is about point-two, which means we're recording only about one Cooper's hawk for every five present in a given area. The only way you can see hawks in numbers and get a sense of their true numbers is on the migration counts, and those counts prove there are tens of thousands of raptors going undetected by other censusing techniques. The migration data are long-term, also. You can't talk about population trends in raptors unless you are looking at five or ten years of data and preferably twenty or thirty years. Hawk Mountain has taken a lot of heat from ornithologists over the years, but the data set from Hawk Mountain is fifty years old now and better than anything else we have."

One last value of hawk lookout counts has nothing to do with their academic reputation, the size of the data base they have generated, or their potential as an environmental early-warning system. Yet it may be the most important value of all. It is a dividend which is widely overlooked by hawkwatchers themselves, especially when they are defending themselves against their academic critics. Hawk-migration counts can be justified scientifically, but they are far more important as educational and political instruments.

The DDT story hawkwatchers tell so often to justify their existence is, in many ways, a misleading tale. The plot seems perfect, of course: how a chemical pesticide created a crisis, wiping out the peregrine falcon in the East and devastating raptor and other animal populations everywhere; and how the conservation community, hearing the outcries from birders and other naturalists, united against the poison and pushed for the ban that eventually came. The thirty-year episode was a tragedy with an upbeat ending. Chemical science made a mistake, environmental science detected it, and though the peregrine may never return to nest on the cliff sides it once occupied throughout the East, the osprey, the bald eagle, and other species were saved and are now recovering. The hawkwatchers' version of this story invariably emphasizes how DDT spread through the environment before anyone recognized the threat it posed. We learned a lesson there, they'll tell you: we have to be ever-watchful. And that's why migration counting is so necessary, they say—as a protection against future DDT-like catastrophes.

But DDT was an anomaly. It came upon us like an enemy invasion—sneaking and swift, destroying so widely and terribly that scientists, naturalists, and ordinary citizens leapt to beat it back—*and* it could be beaten back. DDT was banned, a few wealthy chemical companies lost money, the rest of us lost nothing, all of us benefited by the removal of a health hazard, and the natural world recovered. This is not a common plot in environmental stories, unfortunately. Most environmental problems are so much less dramatic that they cannot be called crises and rarely make the newspapers. The most significant are far more complicated than DDT and have no easy answers. The unchecked growth of human habitations and consequent loss of open areas all along the Atlantic Coast is one such problem; the continued fragmentation of woodland forests throughout North America is another; a third, one that does make the newspapers, is the inexorable buildup of carbon dioxide and other waste gases in the atmosphere, the Greenhouse Effect. None of these problems will take us by surprise. None has a simple solution. None will go away quickly. In fifty or a hundred years conservationists will still be struggling with the destruction generated by these developments and will look back upon DDT in the way we look back today on the feathered hats of the turn of the century: as a minor annoyance created

Kestrel over the dunes

by a temporary style, a problem that virtually solved itself. They may also look back on the last decades of the twentieth century as we look back on the last decades of the nineteenth—as the final years of open space and wild lands.

Coastline development, forest fragmentation, and the Greenhouse Effect are problems that *cannot* be solved, at least in the foreseeable future. The best we can hope for is to check and slow the destruction. Unlike DDT and feathered hats, none of the current threats develops from the misguided preferences of a minority of citizens, and none can be addressed without great cost to all of us. And finally, each has developed so slowly and so widely that the destruction caused seems inevitable. We see another housing development go up on beachfront property, and we groan and walk away with a shrug. What can anyone do? Nothing, we think, nothing at all.

Conservationists, like the rest of the public, respond best to minor dramas and bite-sized crises and not well at all to large, long-term problems. The capture of the last free-flying California condor makes the evening news; the decade-long decline of the sharp-shinned hawk does not. Birders and other naturalists contribute to this syndrome by their emphasis on the exotic and the unusual, and by their love for endangered species. It is a habit we must break. "If uncommon is good and rare is better," Pete Dunne has written, "what does that make extinct?"

"The time to save a species," Rosalie Edge of Hawk Mountain preached, "is when it is still common." She was right, of course, but to save a species when it is still numerous we need to celebrate its abundance—to make a common species seem as exciting as a rarity—and convince more people to care about it. We must better appreciate what we have now, while we have it.

This is what hawk counts do far, far better than any raptor research project: they celebrate the common abundance of hawks and draw more and more ordinary citizens to take note of that abundance.

Hawkwatching is an acquired taste and one that does not come quickly to those who lack companions and mentors. It is not easy for a beginning birder to step outside with a pair of binoculars and enjoy a passing migrant. The birds fly high and fast; most are hard to identify. But join the crowd on the Cape May Point platform (or at any other hawkwatch) for an hour or two on an autumn day and it's hard not to get caught up in the fun of it all, no matter what your level of expertise. Take a look at the Cape May Totals Board and you can't help shaking your head: where do all those hawks come from? Talk to Clay Sutton or Jeff Bouton for a couple of hours and you can't help hoping they will see more hawks this year than they did last year. Spend three or four days watching hawks from the Point's platform this fall, and you may very well find yourself next July paging through your calendar, planning how you can do the same thing again.

"Hey," Bouton announces, looking up from his clipboard after his noontime tabulation. "We're already over a thousand hawks for the day. This is the earliest we've made it to a thousand all season."

A couple of the watchers standing nearby nod and smile. The rest seem not to think that a thousand hawks in one Cape May autumn morning is an event worth remarking upon. And, in a sense, it isn't. Every Cape May autumn there are ten or twenty days when the count breaks a thousand by noon, and they are usually mornings just like today—without rarities or huge kettles, when the birds come by so steadily they seem to be riding conveyor belts. Their flight is so natural and consistent that it's hard to remember the mornings when the sky is empty of hawks. Bouton likes to tell new visitors about the Hawk-o-Matic, a giant fan, five stories high, that he says stands just out of sight over the horizon, north of Cape May City. On dull days he will pretend to make a fumbling search for the button somewhere along the platform's railing, pushing at a knot in the wood and looking up, "Well, I guess that's not it," and pushing at another knot. Today there is no need for his routine. "Today," he says, "the Hawk-o-Matic is in full operation." He walks to the edge of the platform and calls over to Sutton. "Hey, Clay, we just broke a thousand."

Sutton lies in the grass near the bicycle rack, surrounded by half a dozen other watchers, including his wife, Pat, all of them flat on their backs and scanning the sky. He raises one hand and gives Bouton a semi-salute.

The supine position is Pat Sutton's idea. She lay down to rest after a picnic lunch one recent afternoon and spotted a dozen hawks in a couple of minutes, all of them at the zenith of the sky, riding the noontime thermals to the limit of vision and floating by undetected by other watchers. Today, she and Clay and their companions are aiming their binoculars straight up and spotting hawks so distant they could be mistaken for dragonflies. "There's an osprey. See it by the Y-shaped cloud?" "Yeah. And there's a sharpie just above it." "It's two sharpies, isn't it?" "You're right." "Hey, everybody, here's a adult Cooper's." "I've got a little kettle of broad-wings over here, five of them." "Oh, neat." "Two more ospreys right here."

"This is great," says Clay. "This is going to be our answer to the noontime lull."

Cape May is at its best on a day like today. Summer is over, winter is coming, the wind is easy from the northwest, and the hawks are going south. Watch them come for an hour or two and it becomes hard to talk, or even think, of anything else.

"*Adult male harrier* down on the deck," says Pat Sutton, "low over the dunes." She has seen thousands of harriers, as has each of the other watchers, but she and the others twist and roll on their bellies to watch it go. The adult male harrier is the prettiest raptor in the East, with a gray hood and underwings as white as bed sheets on a clothesline. *The gray ghost,* banders call it. "Isn't that a *beautiful* bird?" Pat asks.

No one answers her. No one has to. Here comes another.

Chapter Seven

❖

FISHING IN
THE SKY

THE ARRANGEMENT OF NETS, TRAPS, LURE BIRDS, AND BLIND AT THE FAR
North Banding Station forms an inner diamond and outer semicircle
much like a baseball field's. The line of trees at the end of the open
field, where the grass grows high, curves like an outfield wall. At
home plate is the blind, a dilapidated eight- by twelve-foot shack that
might be mistaken for an abandoned hotdog stand. Signs are stapled
to both side doors:

NOTICE!

These traps or nets are operated by licensed cooperators of
the U.S. Department of Interior's Fish and Wildlife Service. The birds are caught alive, marked with serially
numbered metal bands, and released unharmed. Interfer-
ence with the equipment and operation of the station is
unlawful.

U.S. Department of Interior
Fish and Wildlife Service

Fishing in the Sky

❖

Mist nets run diagonally outward from the two front corners of the blind, angled like the fencing of a baseball backstop, though they are panels of very fine, black nylon mesh about eight feet high and thirty feet long, strung between metal poles like oversized badminton nets. These are the station's passive traps. Hawks are caught when they swoop low across the field, strike the nearly invisible netting, and drop into the pockets of mesh at the base of each panel.

The more effective traps are positioned in the short grass between the mist nets, like a pitcher, four infielders, and a runner on second. A lure bird sits at each, connected by a leather harness and a rope that leads through a pulley and ceramic connectors back to the blind. A second rope runs to the blind from the straw behind each bird. Hidden beneath the straw at each trap is a spring-powered hoop net.

Overhead lines of cormorants, chevrons of geese, and whirling circles of tree swallows are moving south in a blue, cloudless sky. "Look at all those birds," says Joey Mason. "It's going to be a good day." She is sitting on a foam rubber pad on a wooden bench inside the blind. The traps' twelve lines are at her fingertips.

A harrier appears, banking low, well beyond the trees, two hundred yards away from the blind. "Okay, pidge," says Mason, "out you come." She pulls on the line leading to the trap where the pitcher's mound would be, and the lure bird there, a white pigeon, flutters into the air. This trap, the bander's "main bow," has a ten-foot-high pole with a pulley on top. The pigeon's line leads up through the pulley back to the blind. Mason lifts the bird as she yanks on the line, and the bird flaps awkwardly in response, looking like weakened and promising prey.

The harrier ignores the bait, however, and drops over the horizon.

Mason wears a ski cap, sunglasses, and three sets of earrings—two gold hoops and a cluster of blue beads. Sitting on the edge of

Far North Station

her seat, leaning and rolling forward to look right and left out the sides of the blind, then leaning backward to glance through the slit in the roof, she seems as alert as a child on a tram ride through an amusement park. The two-way radio on the floor between her feet links her with Jeff Bouton on the hawkwatch platform and with the three other banding stations in operation this morning—East, North, and Hidden Valley.

Ordinarily, banders work in pairs in each blind, but today Mason is by herself. "I prefer it this way. It's real peaceful, and when you catch something good, it's a lot more exciting. You just have to hope you don't get footed." To be footed is to be grabbed by a hawk's talons, squeezed in its sharp grip. Banders say being footed by a kestrel or a sharp-shin feels like having your arm impaled by a crocheting needle or pinched by a pair of pliers tightened down hard. Asked how it feels to be footed by a larger hawk, they suck in air and shake their heads, eyes closed. Among the unwritten laws of the Cape May banders is the maxim that you never let go of the hawk, no matter how badly you are footed. You hang on to the bird while your partner pulls out the talons. If you are alone in the blind, you hang on with one hand and pull out the talons with the other.

A sharp-shinned hawk flap-flap-glides into sight, swings low over the left-hand bow trap, then flaps upward and clears the mist net on that side by a couple of inches. Five minutes later, another sharp-shin repeats the sequence.

"Oh, they're seeing the nets this morning," says Mason. "It's probably the dew in the mesh. I'm going to have to depend on the lure birds until the sun dries everything up."

A moment later, a third sharpie lands on a pole beyond the mist net.

"I've got to run my sleaze out," says Mason. She tugs on a line, and the starling at third base emerges from a cavity in the straw and dirt twenty feet from the hawk. Starlings are "sleazes" because they are such slovenly birds. "Come on, sharpie," Mason whispers. "Come to the smorgasbord."

The hawk drops down, landing next to the starling with a quick flick of its tail. It looks around, hops a step closer—*thump!* the bow slams the ground. The sharpie is trapped in the net. It jumps weakly upward, too tightly tangled to move its wings. Mason runs out, slides her hand under the bow, grabs the hawk's legs, and lifts it out. She resets the bow with her other hand, replaces the straw, and runs back inside, holding the hawk.

From the collection of containers stacked on the work table on the blind's back wall Mason selects a pair of V8 juice cans taped together end to end and punctured at the far end with half a dozen air holes, and she slides the hawk inside, headfirst. "It's a little male," she says. "Females fit in the Pringles can." The shelves above the table hold a toolbox, two sets of pliers, a can of WD 40, an AM/FM radio, a six-pack of diet soda, a whisk broom, a plastic ruler, a set of calipers, a clipboard thick with forms, several field guides, and a balance scale. Eleven strings of bird bands, fitted on wire loops by size, hang like silver necklaces from nails below the shelves.

As Mason choose a band, size 2, from one of the loops, the hawk seems catatonic, its toes dead still. Most hawks seem to remain remarkably calm while they are being processed—staring upward at the bander as calmly as a hospital patient watching a nurse take his pulse. Even the few hawks that are feisty after capture grow quiet and still once inside the holding container. "If the can fits snug," says Mason, "they won't move around. Your first thought is give them a little room, but actually they're better off when it's tight. They can't bung up their shoulders then. Once in a while they might bite at the can and make a little noise, but they can't hurt themselves."

She reaches for the smaller set of pliers and, using two prongs

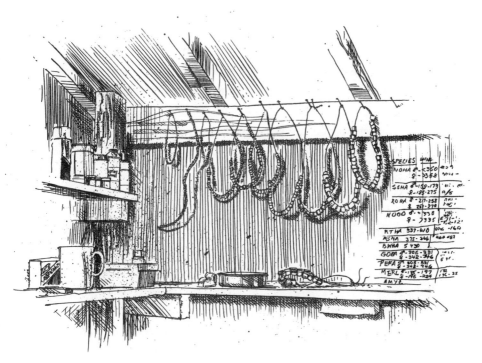

"Bird bands hang like silver necklaces . . . "

at the pliers' tips, levers open the two halves of the band. She glances up through the roof's slit and over her shoulder at the traps behind her as she works.

"Freeze!" she whispers. A hawk is standing on the white pigeon at the main bow. "That's a Cooper's hawk, and it's an adult, I believe." She places the pliers softly on the table, takes one stride to the line to the main bow, and yanks. *Thwack!* the bow thumps the ground and the Coop is caught.

She runs out, reaches under the bow to grab the hawk by the legs in one hand, and holds it shoulder high while she pulls back the bow with her other hand to reset the trigger, covers the trap with straw, checks the harness on the lure bird, and runs back to the blind.

The hawk is an adult male Cooper's, a third longer than the sharpie. In profile, the dark, down-swept cap, pushed-in bill, and flaming orange eye give it the look of a junior high school tough. It glares back at Mason as she slips it into a can that once held Hawaiian Punch. The legs stick out under the long tail, as thick as pencils. The feet and talons seem twice the size of the sharpie's. "These are feet you respect," says Mason. Like several others of the Point's banders, Mason prefers to work bare-handed. "Gloves get in the way," she says. "You can't feel the bird."

She selects a band from a different loop—size 4—opens it with the pliers, slips it on, and squeezes gently. "You always put the band on first. That way, if the bird escapes during processing, at least you've gotten the band on. That's the most important thing. That's why we're here." She rolls the band between her fingers to be certain it is not too tight, then slides the hawk out of the container. Gripping the bird around its legs with her left hand, she measures the bill with the calipers with her right hand, puts down the calipers to pencil in the number on her clipboard, puts down the pencil and picks up the calipers to measure the talons, pencils in that number, puts down the pencil, picks up her ruler to measure wing length and width, pencils in those numbers, fans the tail and wings to count and examine each feather for molt, and finally pencils in those details on her clipboard.

She slides the bird back inside the container to weigh it on the balance scale: 347 grams, three-quarters of a pound. Then she slides it out again, opens the side door, and releases the bird into the air. "Bye-bye, sweetheart."

The sharp-shin has remained unmoving in its container. Mason quickly presses its band together, checks the tightness, slides the hawk out of the container, and releases it out the window. Cape May's banding operation has caught so many sharp-shins over the years that the banders no longer bother to weigh and measure them.

Sharpies and kestrels, the next most common capture, are simply banded and released, "rung and flung" in banders' jargon.

Mason has a flurry of mist net captures in the next thirty minutes: sharpie, sharpie, kestrel, sharpie, sharpie. She processes each quickly, ringing and flinging.

"I know some of the local birders don't like the banding project, but I just don't understand why. You can't study migration unless you can track individual birds, and you can't track individual birds unless you band them. What else can you do? Some projects use color dyes, but you can't mark birds individually with dyes, and as soon as the bird molts a couple of months later, the dye is gone. A band stays on forever. You can recover a peregrine ten years after it was captured, and the numbers on the band tell you exactly where it came from, how old it is, and who caught it last time.

"Humane-iacs don't like banding because they worry about the lure birds. They think we're torturing them. It's just not true. We take excellent care of our lure birds. Every morning as soon as I get here I go out back to talk to the lure birds, clean their cages, change their water, and feed them. The last thing you want as a bander is a lure bird that won't do anything for you."

Two more Cooper's cross overhead. "Come on, guys," Mason says, yanking on a line. The white pigeon flutters into the air. "Come on down to Joey's Bar and Grill."

Joey Mason is thirty-five years old. This is her fourth year banding raptors at Cape May Point. "My brother Austin is the real birder in our family. He finds a feather in the road and he knows what it is. I wasn't interested in birds until the year after I graduated from high school. I was working at an animal hospital in Massachusetts, and a friend and I drove down the East Coast together in my mom's Mercury Cougar. We made it all the way to the Everglades. We stopped at Mrazek Pond, and I took a photo of the reflection of a

Louisiana heron. The shutter setting was wrong, but that picture came out great, a real moody green. I used to look at it all the time. I still have it.

"I went back to the Everglades a couple of years later and shot rolls and rolls of film. Mostly long-legged waders—roseate spoonbills, white ibis, great egrets. I just love those birds.

"After college I worked in a couple of animal hospitals, a camera store, and then in a flower shop for six years. I bought a copy of George Harrison's book, *Roger Tory Peterson's Dozen Birding Hot Spots,* and decided every vacation I was going to try to get to one of those twelve places. The first year I drove up to Machias Island in Maine to see the puffins, and the next year I went to Hawk Mountain. A couple of thousand broad-wings came by my first day there. That put a crimp in my plans to go to the rest of Harrison's spots. I went back to Hawk Mountain every September for eight years in a row—just to sit out on the rocks and watch those broad-wing flights.

"Harrison's book talked about Cape May and the hawk migration over the Point, but I never wanted to go. The whole state of New Jersey was such a pit as far as I was concerned, I couldn't imagine driving through it. Then in nineteen eighty-five, Jim Brett at Hawk Mountain told me I ought to go see Pete Dunne at Cape May about working as a hawk counter. So I finally drove into New Jersey. Pete said he didn't need another counter, but maybe the banding project could use some help. 'You have to go ask Bill Clark,' he said.

"I met Bill on the platform and he took me over to South Station. There were nets everywhere and lines crisscrossing back and forth. 'Don't touch *this,*' he kept saying. 'And don't touch *that.*' It was pretty intimidating. I'd never even seen a blind before, and I was terrified I'd trip on a line and pull over something.

"What I remember best is all the Pringles and V8 cans against the wall with tails of hawks sticking out. There were three banders in the station—two people trapping and Chris Schultz, who was processing about eight birds at once. 'Whew,' I said, 'what are these?' 'That's a sharpie,' Chris said pointing. 'That's a kestrel. That's another sharpie. That's a merlin.' 'You can tell just from the tails?' I said. 'Oh, yeah,' he said.

"That was September twenty-sixth, nineteen eighty-five. Hurricane Gloria hit the next morning. When the rain finally stopped, the whole Point was about a foot underwater. I helped Chris rebuild the blind in East Station, which had been swept up onto the dunes. In return, he taught me banding. I stayed for five weeks.

"The first bird I caught was a kestrel. Chris sat at my shoulder and talked me through it. The second bird I caught was a harrier. The third was a peregrine. He talked me through each one—'OK, get centered . . . hold it . . . ready . . . *now!*' I was hooked, to say the least.

"Toward the end of that year he would let me stay in the blind when he had to go back to the park to do a demo. Which scared me to death. I just prayed nothing would come.

"The next fall Chris got me a sub-banding permit, and he let me stay alone a lot. I caught my first gray ghost, my first red-tail, my first goshawk. The first goshawk I ever saw in my life was one Chris caught. He was at another station, and he called me on the radio. 'Hey, Joey, you want to see a gos?' I closed down my station, and ran over to his to look at the bird. Then I came back to my station, and an hour later I caught my own gos. My hands were shaking, my heart was beating—hold it! Red-tail."

A red-tailed hawk circles directly overhead, its fanned rectrices the color of crushed strawberries.

Mason yanks the line to the white pigeon, and the hawk

responds immediately, braking in the air and flapping hard to hold its spot in the air.

"Isn't that incredible? It's hovering for me, not just over someone's field. It's responding to *me*. People say banding is like fishing in the sky. I say this is a lot better than fishing."

The hawk hovers for half a minute, looking hard, then returns to its soar, circling up high into the blue. Mason unbuckles her wristwatch, "I think it's reflecting in the sun. A red-tail can see something like that from a mile away." As she hangs the watch from a nail above the window, a sharp-shin lands on the sparrow at the trap to the right of Mason's main lure. *"Shit,* it never fails. You've got something good overhead and you catch a sharpie." She flips the bow trap over the hawk and runs out. The red-tail floats off and away.

"What a glutton you are," she says as she carries the sharpie back to the blind. "You have a full crop and you want more?" She strokes the bird's throat which shows a bulge the size of a ping-pong ball. "What have you got in there, sweetheart—some little warbler?"

The radio crackles, Jeff Bouton's voice: "Hawkwatch to North. You've got a golden eagle coming your way. . . . He's over you . . . right . . . *now."*

Mason leans forward to peer over her right shoulder through the roof slit in the direction of North Station. "If North's got it now, we'll get it next. *God,* I'd love to catch an eagle. No one's caught one in years, but when one comes over your station you *pop that pidge.* If the eagle turns its head to look your way, you don't even breathe. If you have to blink, you keep it short."

The eagle appears a minute later, low over the trees beneath a kettle of turkey vultures, perhaps three hundred yards away. It banks once, circling up to join the vultures, and white wing patches flash—it's an immature. Mason pops her pidge furiously. The line

whips and whistles across the connector. The pigeon is lifted up and
flutters down flapping, is lifted again and flutters down—up, down,
up, down. The eagle never turns its head. In a couple of minutes it
is a dot in the blue over Cape May City.

And, thirty yards away, a Cooper's hawk is standing on
Mason's pigeon.

"It must have come from the woods behind us," she whispers.
"Cooper's are sneaky that way. . . . Come on, get your wings
still—oops, oops, here comes that red-tail."

The red-tailed hawk swoops down, legs dropping like a
jetliner's landing gear, and flushes the much smaller Cooper's back
into the woods.

The red-tail back-flaps once after landing, steps leisurely onto
the pigeon, and looks around. Mason reaches for her trigger line and
waits. The moment before the trapper pulls the trigger is the most
critical moment in the sequence. The bander must be certain the lure
bird is "centered"—close to the post that sits in the middle of the
bow's throw space. The spring that powers the bow is large and tight;
the hoop flies faster than the eye can follow. If the hoop hits the bird
it will shatter its wing, and the larger the bird the smaller the margin
for error.

The red-tail flaps once more to hold its balance, closes its
wings, and Mason triggers the bow. *Thump!*—the hoop whacks the
ground. The hawk hops against the net, tangling itself in the mesh.
It's a clean capture.

Mason sprints out, reaches under the net to grab the hawk's
ankles, and lifts the huge bird in one hand. It flaps rapidly half a dozen
times while Mason resets the trap, covers the hoop with straw, and
checks her pigeon. She strides back to the blind carrying the red-tail
shoulder high. "After I process this one, I'm going to switch lure
birds. Give that pidge a break."

Two twenty-four-ounce coffee cans attached end to end with black adhesive tape are the hawk's holding container. Its yellow feet protrude, as wrinkled and thick as an old man's knuckles. Mason measures one leg with a gauge, then clips on a size 7B band with pliers. She slides the bird out of the container to measure bill and talons and study each primary and tail feather. She notes all details on her clipboard, then pushes the hawk back in the container to weigh it on the balance scale: 1,007 grams. The hawk, which is as tall as a cat and has a wingspan wider than a child's spread arms, weighs barely two pounds.

"Yukko, a flat fly." An insect walks up Mason's forearm. A *hippoboscid,* a parasite that feeds on the blood of birds, it is ink-black and as laterally compressed as a Stealth Bomber, with pincer claws on its front legs designed to enable it hold on to the feathers of its host even in a top-speed dive. "They can't hurt you, but it's pretty gross when you go into the Acme and one flies out of your hair and down the aisle."

When Mason releases the red-tail, it flies only a hundred yards, to preen in a dead tree on the edge of the field. It stays there while she replaces the lure pigeon with a new pigeon from the cages behind the blind, and it is still there fifteen minutes later, rubbing its bill across its breast, when Mason spots a harrier a couple of hundred yards beyond.

She pops her fresh pidge; the harrier tilts and dives. "It's coming, it's coming, it's coming. God, look at that stoop!" The harrier lands next to the pigeon: Mason waits a long moment— "Come on, set your wings"—then triggers the bow. *Thump!*

It's a tawny-gold female, half as big as the red-tail and a size-5 band. "Oh, you're my favorite," Mason tells the bird, whose eyes are wide and caramel brown. "Hi, sweetheart." Mason finds a broken feather in the tail. "This must have been hurt before. That was a perfect capture." She notes the feather's number on her data sheet, then slides the harrier back in its container and cuts away the broken tip with scissors.

"Get back in your hutch, sleazo!" she shouts looking over her shoulder. "What are you doing out there?" An accipiter pounces on the starling in the left-hand trap. "Oh, no, another damn Cooper's, right on my sleazo." She places the harrier back on the work table, steps forward, and yanks on a trigger line. *Thump!* The hawk tumbles over, trapped.

Mason releases the harrier, then takes the Cooper's out of the net. The starling is apparently unhurt and retreats into its dugout again. Mason is examining the Cooper's wing feathers, after banding the bird, when she stops and lifts the hawk to her nose. "You have an odor." She studies the banded leg and finds a fine white line, apparently a wound that healed long ago. Mason doesn't want to take chances. "If it started to swell, the band could be a problem, so I'm going to switch legs. You have to remember, any bird you trap is going to be wearing your band for the rest of its life, so you have to put it on as carefully as possible, whether it's a sharp-shin or a peregrine."

Mason rummages through the toolbox and takes out something that looks like a combination of gardening shears and needle-nosed pliers. "I forget what they're called. I found them in an auto supply store and then filed them down. Getting a band off is always a bitch, but these work pretty good." She pokes the tool under the band and wedges it open. Then she bands the bird on the other leg, finishes the processing, and releases the bird out the window.

It is 12:58 when she stands to check her watch hanging from the nail above her. She's been too busy for lunch. The first red-shouldered hawk of the day is soaring overhead, wing crescents twinkling in the sunlight, and a sharpie bangs into one of the mist nets. "Boing!" says Mason. "Another sharpie. Sometimes you hope they bounce out. Uh-oh, it's wearing a band. It's probably a bird I caught this morning."

It's an immature female, and she slips it into a Pringles can. An inch-long, bright blue feather clings to one talon, remnants of the hawk's last meal. "No, this isn't my bird," says Mason. "That's not my band." She picks up her radio, "Far North to East."

Chris Schultz answers, "This is East."

Inside the blind

Mason rolls the band in her fingers. "Sharp-shin 1043-52077."

Schultz calls back in five minutes. "East to Far North. That sharpie was banded last week by Bob Yunick. It seems to be hanging around."

In North America in recent decades, bird banding, like hawk counting, has become primarily an enterprise conducted by amateur ornithologists. The long hours and low rate of returns seem to discourage most professional biologists. The odds against recovering any banded bird, especially any migratory species, are steep. In 1965 a study of all banding records of the white-crowned sparrow revealed that only 198 of the 266,516 white-crowns that had been banded over the last forty years had been recovered at any distance from the original trap. This is a typical recovery rate for migratory songbirds. Because their corpses more often draw the attention of passersby, the rate of recoveries for banded hawks is slightly better, between two and five percent, but it is still so low that very few professional ornithologists stake their reputations on their work as banders. The work is left to amateurs like Mason and the other Cape May banders, who seem to find their satisfaction in the daily pleasure of handling the birds.

Mason is paid $150 a month to band eighty-five hours a week—about forty cents an hour. Her room, at the Choate House on York Avenue where the banders stay, is paid for. Her board is not. At the moment she has no other job. Last summer she worked censusing and studying falcons under Schultz on the peregrine project in Colorado. She's hoping to do it again in April. She is not sure where she will spend the winter. "One thing I know is I'm not sticking around here as long as I did last fall. I was still banding Thanksgiving Day. This year I promised myself I'm not staying past the first week in November, no matter how much I want to."

* * *

On the hawkwatch platform an hour later, Jeff Bouton, who over-heard Mason's exchange with Schultz on the radio, is talking about the retrapped sharp-shin. "When I hear about things like that, I start wondering what the hell I'm doing here. That sharpie has been around the Point for *two weeks.* Are any of these so-called migrants really going anywhere? Maybe all we're doing on this hawk count is what all the nonbelievers say we're doing—counting a lot of the same birds over and over.

"Look at the recovery records: a red-tail trapped here found the next week in Media, Pennsylvania; a sharp-shin trapped here picked up dead three days later in Westchester, Pennsylvania. Last year I saw a peregrine heading down the beach with orange dye on its breast and a radio antenna sticking out its rear end. That bird had already been down to Virginia and all the way back again in two days—the banding project on Assateague Island had trapped it. A couple of weeks ago I had a kestrel with two albinistic primaries fly by me eight times in one day and three times the next day. I wouldn't have known it was the same bird if it hadn't had those two white feathers.

"Under Kerlinger's new rule, I couldn't come back as the counter next year even if I wanted to. He says that from now on nobody will be allowed to be the official counter for more than two years in a row. He says he has to do that to minimize observer bias, quote unquote." Bouton shrugs. "I don't really give a damn what his reasons are: I wouldn't do this count again next year if they begged me. It's too damn frustrating. You can't tell what the birds are doing.

"Maybe I'll come back next year as a bander. I took a day off for mental health purposes a couple of days ago—slept in, wrote some letters, took care of some other things pressing on my brain. Then I came over for the demo, ran into Schultzie, and asked him if I could visit his blind. 'Sure,' he said, and later he let me catch a male harrier, a female harrier, a male merlin, and a couple of sharpies.

Wow, that felt good. People say, 'Those damn banders, they're only interested in banding because it's fun.' *Of course* it's fun! Any bander who says it isn't is a liar."

Pete Dunne and Clay Sutton are sitting on the bench behind Bouton planning the raptor program they're presenting tonight in Cape May. It's a slide show they have reorganized half a dozen times since the publication of *Hawks in Flight,* the book they wrote with David Sibley. Their presentation has the universal problem of all bird-identification programs—the lights go out, the details grow heavy, and the least interested viewers fall asleep—and another, more complicated problem. The system of hawk identification they are advocating, the "holistic" method, differs from the traditional method principally in its emphasis on subjective impression and in-flight behavior. It enables the watcher to identify birds at much longer distances than the traditional system, but requires more experience and demands more judgment. Standard field guides emphasize definitive field marks: the crescent-shaped slice of white, the "window," in the wings of a buteo, for example, means the bird is a red-shouldered hawk; a falcon with red tail must be a kestrel; and so on. Even a beginner can identify a hawk by such field marks—but the bird must be close and the light just right. The field marks described in *Hawks in Flight* can be used at longer range and poorer light than the standard field marks, but they are more subtle and less definitive: "The [red-shoulder's] wings are long, narrow, and clean-edged," Dunne, Sibley, and Sutton observe, "and lack the muscular bulges seen on the red-tailed. The leading edge is straight. The trailing edge curves gently (usually on adults) or not at all (usually on immatures). The wing tip is cut straight on an angle. Seen from below, the wing of the red-shouldered suggests a long, rectangular plank [and] juts forward when the bird is in full soar, as if it were reaching out, arms wide, to embrace something."

Learning to identify hawks by the Dunne-Sibley-Sutton

method rather than by watching for the standard field marks is like learning to play chess by positional strategy rather than by watching for simple two- and three-move combinations: you must be experienced enough to have developed an intuitive sense of the game, motivated enough to study numerous complexities, and knowledgeable enough to understand the method's limitations. For birders at the level of ability between intermediate and expert, *Hawks in Flight* is a wonderful collection and distillation of hawkwatching lore. For beginners, however, it is generally too much. Since the audience at most slide shows is a heterogeneous mix of birders at different levels of ability, from *patzer* to grand master, Dunne and Sutton are struggling to find a way to engage the interest of the beginner, present just enough of their method to satisfy the intermediate birder, and avoid provoking any of the experts in the crowd into contradicting them.

Complicating the matter still more is the fact that the subjectivity at the heart of their method leads Sutton and Dunne into conflicts with one another. Sutton, the photographer, has a more geometric eye: "Kestrels are all soft angles," he likes to say. "Merlins are more hard-edged." Dunne, the wit and the wordsmith, prefers metaphors and similes, and the farther he must stretch (and the more distressed Sutton becomes) the happier Dunne seems to be. "Kestrels," he will pronounce while Sutton squirms, "have banana wings. Merlins are stocky and blunt, the Porsche Carreras of the falcon clan." Sometimes the coauthors find themselves pooh-poohing each other's descriptions in front of their audience.

Tonight, Dunne wants to try a different approach. "Let's just show the slides and let them tell *us* what each one is. Then we can ask, 'How do you know?' That way it becomes more of a back-and-forth discussion between us and the audience."

Sutton shakes his head. *"No way.* We'd have to limit the

whole show to about eight slides, and we'd run out of energy a lot sooner than they would."

"So what? How many slides can the average beginner absorb? We show them your four hundred slides and they OD."

Dunne has touched a sore point here, and he knows it. Sutton has earned an unwanted reputation for presenting the longest slide shows of any birder in New Jersey—sometimes four trays of slides long. While heads bang on chair backs all around the room, Sutton keeps going, slide after slide, as if he can't help himself, like the overenthusiastic museum guide so hopeful that one more picture might convince someone else to share his love for art that he can't bring himself to end the tour.

Sutton stares at his friend. "The whole show isn't even three hundred slides, Pete."

Meanwhile, the other birders on the platform are debating the identifications of two mystery birds of the past week: a dark-legged, brown-backed sandpiper seen by a dozen observers at a tidal pond in Stone Harbor and believed by some to be a Eurasian little stint; and an odd-looking cormorant that flew by the platform in a flock of double-cresteds. More than twenty veterans watched the cormorant pump past, but no one seems certain about its identity. Some say it was probably just an underdeveloped, "runt" double-crested; others say it might have been an olivaceous cormorant, a Mexican and South Texas species that has never been recorded in New Jersey.

"I think olivaceous is a good possibility," says Bouton. "It was *dinky*—it seemed about the size of one wing of the double-crested it was flying next to."

"That would make it a runt olivaceous, Jeff," says Bob Barber. "What's the length of one wing of a double-crested, eighteen inches?"

"I said it *seemed* that size. That was the impression I got. I wish I'd had my camera. . . . I did photograph the little stint, thirty-six shots at about ten yards."

"The *hypothetical* little stint."

"That's what I called it until I saw it," says Bouton. "That bird just jumps out at you. The breast is dark, the back is a real warm brown. The instant you see it you'll know it's something different. It's probably still there."

"Forget it. Nobody's reported it in the last five days— nobody who knows what they're doing."

"Nobody who knows what they're doing has *looked* for it in the last five days."

"A sandpiper isn't going to hang around in New Jersey for two weeks in the middle of October. A little stint should be in Africa by this time of year."

"That bird doesn't know where Africa *is,*" says Bouton. "He's totally lost. Maybe he's sticking around here because he doesn't know how to get to anywhere better."

At 5:30 Bouton closes down the hawkwatch and leads an expedition of eight birders in three cars up the Garden State Parkway to Stone Harbor to search for the stint. He pulls over in the New Jersey Wetlands Institute parking lot. The sun has begun to set, and the only birds in sight are two great blue herons a hundred yards away and a dozen unidentifiable gulls even farther off. "This is where it was first seen," he says, pointing at one end of a bird-less tidal pond. "And that's where I saw it," he says, pointing at the other end. "Let's try across the road."

The group crosses the road, through heavy, sixty-miles-per-hour traffic. Not a single bird is in sight. They sit down, one after another, on a discarded telephone pole until only Bouton is left standing. Finally, he wanders far out onto the lifeless flats, which are littered with beer cans, truck tires, and plastic bottles. He returns with a Styrofoam life preserver on his head, tilted like a sombrero, and looks at the row of dispirited birders. "Jesus, what a waste of time. Whose idea was this, anyway?"

At 7:30 an audience of a hundred assembles at the West Cape May Municipal Court, the space borrowed by CMBO for Dunne and Sutton's raptor program. Most of the banders and most of the people who spent the day on the platform are there, as are dozens of others carrying Dunne, Sibley, and Sutton's book. Its royal blue cover is in every corner of the room.

Sutton leans against a far wall while Dunne whips the projector cord free from the front row and steps forward. "Clay and I have been accused of talking raptors far past the point where anyone else wants to listen, so tonight we're promising to restrain ourselves. Check your watches. We're limiting this talk to one hour."

Click. The first slide shows the ridges and farmlands below Hawk Mountain. "Hawk Mountain is the Godfather site," says Dunne, "the mecca of North American hawkwatching, the place where it all got started. I strongly recommend every single person in this room make the pilgrimage to the Kittatinny Mountains of Pennsylvania to stand on the North Lookout at Hawk Mountain. That way you can see for yourself . . ."—*Click,* a kettle of sharp-shins over the Point's platform—"how good Cape May is!"

Click, click, click. A series of shots of distant hawks. "Open up any of the standard field guides and you'll see all the hawks are drawn at close range in bright colors from what seems to be about ten feet away. Anyone who has ever spent half an hour outside actually looking at hawks knows you don't see too many from ten feet away. Hawks tend to stay off in the distance, where field marks like throat color and tail-band width don't work very well. We wrote our book because over the last few decades hawkwatchers around the country have been developing a different system for identifying hawks at long range, using more subtle field marks—tendencies and general impressions rather than single, all-or-nothing clues."

Click. A kestrel flying left with backswept wings. "Clay is going to gag in the corner, but I'm going to say it again: kestrels have banana wings." *Click, click, click:* kestrel, kestrel, kestrel.

Click. A merlin. "Take a kestrel, put it on steroids and a Soviet weight-training program for six months, and what you get is a merlin. The difference between a kestrel and a merlin is the difference between a scooter and a Harley Davidson." *Click, click, click:* A dozen merlin shots, most from long range.

Click. "Peregrine. This is the bird people come to Cape May to see. We're making I.D.'s of peregrines now at distances of two and three miles—from the platform to the Second Avenue jetty. The best clue at long range is their wing stroke: the flapping motion travels down a peregrine's wing like the motion down a garden hose as you shake it."

Click. "Chris Schultz will remember this bird. Right, Chris?"

"Oh, yeah."

"This is a peregrine that terrorized every banding station in Cape May for the whole month of November a couple of years ago."

Click. "Here's the same bird from another angle. The cap is very dark, and Clay thinks this bird might be an *anatum* peregrine. That's the race of peregrine that was extirpated from the eastern United States by DDT and now breeds only in the West. But the dark-capped bird here might also have come from the Cornell Peregrine Project. They breed peregrines in captivity there and release them from hacking sites—peregrines from the *brookei* race are interbred with the *pealii* race, the *peregrinus* race, snapping turtle, Datsun, Toyota . . . no way to tell what one of those mix-and-match Cornell birds is going to look like."

Click. An accipiter. "If you see an accipiter that's clear-breasted and has a hangman's hood like this bird, call it a Cooper's hawk. If you're honest and you're a hawkwatcher—which may be a contradiction in terms—you know you make mistakes. The good thing about hawkwatching is most of your mistakes keep flying away, so nobody can prove you wrong."

By the time Dunne reaches the end of the first tray of slides and turns over the projector control to Sutton, the audience is fading. A little boy, ten years old or so, is pacing back and forth behind the last row. Several adults are slumping in their seats, their faces aimed at the ceiling, their eyes closed.

Click, click, click. Red-tailed hawks in different flight positions and different lighting conditions. "The red-tailed hawk," says Sutton, "is the most variable of all hawks in North America." *Click,* a light red-tail with a dark belly band. *Click,* a light red-tail with a light belly band. *Click,* a dark red-tail with a dark belly band. *Click, click, click:* a series of slides featuring tails—rufous tails, orange tails, pink tails, white tails.

Click. A buteo with a brown-streaked white tail. "This is Harlan's hawk," says Sutton, "now considered a subspecies of the red-tailed hawk. It breeds in Alaska and winters in the prairie states, and if hawkwatchers had a vote, it would still be a separate species. In the field it looks and behaves a lot different than a red-tail. I think of it as a kind of missing link between red-tail and rough-legged hawk. It has the breast and wing colors of a dark red-tail and the long wings and rangy look of a rough-leg."

Click, click, click. Sutton explains how to sex rough-legged hawks: "Immature birds are the darkest. Adult males are next, a little less dark than immatures . . . carpal patches . . . diffuse belly band . . . adult female lighter than adult male . . . progression . . . all very diffuse belly bands . . . patagium. . . ." The ten-year-old boy is asleep, arms around his father's neck.

When the second tray ends and Sutton asks for the third, the groan in the room is loud, but warmed up and talking fast, Sutton seems not to hear it.

Click. "*Look* at this bird! It's a rufous red form of the ferruginous hawk. David Sibley and I found it one winter in Arizona, and David spent about two hours painting it." *Click, click, click:* the red ferruginous soaring, pumping, and banking. "David said it was the most beautiful raptor he'd ever seen in his life."

Dunne takes back the projector control to talk about the northern harrier, his favorite bird.

Click. Immature harrier. "Study this slide and you'll understand the difference between the holistic approach to hawk I.D. and the traditional system. When our book came out, Bill Clark sent us a letter listing what he said were the errors we'd made. One of them had to do with immature harriers. We said they weren't streaked on the breast; he said they were. It's not really a difference of opinion. It's a difference of methodology. Bill's a bander, and when he talks about streaks on harriers he's talking about birds he's holding in his hand. We're talking about how the birds appear in the field at the distances hawkwatchers usually see them. At those distances immature harriers are *not* streaked."

Click. "Here's the classic bald eagle. America's bird, the post office bald eagle, dark breast and white head." *Click.* "Here's what bald eagles look like more often at Cape May. In the immature plumages their heads are dark, their breasts are mottled white, and you might confuse them with this bird." *Click.* "Golden eagle. Golden eagles' heads are always dark, and they have very similar wingbeats to bald eagles. Both eagles have a lot of throw in their wings. The wings go way down and come way back up. The best distinction at long range is the head size. Goldens' heads are much smaller than balds', and they're not as rare at Cape May as people used to think. How many goldens so far this year, Jeff?"

Bouton exhales. None of his totals this season make him proud. "Three," he says in a flat voice.

Sutton consults his watch as he takes over for kites, and he picks up the pace. *Click, click, click, click:* "The black-shouldered kite does everything with its wings above the horizontal. . . . The Mississippi kite never flaps until it absolutely has to. . . . The swallow-tailed kite may be the easiest raptor I.D. in all of North America. . . . " *Click, click, click.*

Click. A shot of Higbee's Woods, glistening green leaves, dark shadows. "We always like to end these talks with this slide and a comment about conservation. A reminder, really. If we want to see hawks, we've got to protect habitat. That's one of the basic principles of CMBO. . . . Now we'd be happy to answer any questions you might have."

The lights come on and several people stand up immediately, blinking and stretching. An older man in the last row zips up his coat and asks, "What trends are you seeing in hawk populations?"

"It's the proverbial mixed bag," says Sutton. "Peregrines, merlins, and ospreys bottomed out during the DDT years and have increased recently. Bald eagles are also doing better. Red-shoulders, on the other hand, have never really picked up after DDT. Their numbers are still down. The species that has hawk counters the most concerned now is sharp-shin. Sharp-shins have shown a steady decline over the last seven years. And kestrel counts are beginning to worry people."

"Any theories about the reasons?" asks a man in front.

"I think the sharp-shin decline is linked to the songbird decline," says Dunne. "We know songbird populations are dropping rapidly. Their nesting grounds are being replaced by shopping malls, their wintering grounds in the tropics are being replaced by cattle ranches, and cowbirds are expanding everywhere." Cowbirds are brood parasites, laying their eggs in the nests of warblers and other small songbirds so their young are raised by the foster parents, whose own young usually die. "All of this means songbird numbers are way down. Anything that eats songbirds is going to be in trouble, and sharp-shins feed primarily on songbirds."

People all around the room are putting on their coats. The ten-year-old boy leads his father out the door. "An *hour?* It was an hour for each tray!"

It's a cold clear night. Jupiter and the Pleiades shine high overhead. Breath smokes. People leave in twos and threes.

Bouton exits pumping his fist. "Chris Schultz offered me the chance to be his assistant in East next year. East is the *peregrine* station!"

Clay Sutton comes out carrying the screen over his shoulder and talking with Jimmy Watson, an old high school friend. They have surf fished together for twenty years, since long before Sutton took up birdwatching. "Yeah," says Sutton, "I want to get out with you real soon. How about Sunday or Monday?"

"Sunday's good for me."

"Great! Let's do it. It's about time. *You can't eat hawks.*"

Chapter Eight

❖

GOLDEN

IN THE 6:00 A.M. PREDAWN TWILIGHT VINCE ELIA DRIVES DOWN NEW England Avenue toward Higbee's Beach in second gear, braking frequently and shaking his head in amazement. He has birded Cape May Point every weekend and most holidays for the last four years and he has never seen anything like this. The birds sweeping back and forth across the road through his headlights look like leaves in a storm. Dozens of flickers and blue jays, hundreds of robins and doves, and thousands of unidentifiable sparrows and other birds are flushing from the roadside bushes, posts, and lawns, and from the road itself. Many refuse to move until the car is within a yard or two, and others swoop past at high speed, veering out of the way inches from his fenders.

Finally, after twenty stops and a dozen near-misses, a flock flushes too late, and the right front fender clicks. Elia stamps on the brakes, *"Goddamnit!"*

He walks around the car, scoops up the bird, and cradling it in one hand, holds it up in the headlight. It's a juvenile white-crowned sparrow. Its head hangs limp, its neck is broken. "Goddamnit."

Elia puts the body gently on the back seat of the car, then drives the last hundred yards to the dirt lot at Higbee's entrance, and swings around to park facing outward, a habit he started in early fall when the lot sometimes held a hundred and fifty cars. Today, October 30, his is car number one, and he knows there will be few people at Higbee's. The warbler peak is over for the season, so most birders will be arriving later in the day and heading for the hawkwatch platform first.

He climbs out of his car, sees another set of headlights coming, and guesses it is Bob Barber's Ford Probe. Barber and Elia bird Cape May so often together they no longer bother to call each other ahead of time. They assume they will meet each Saturday and Sunday here at Higbee's, at the Second Avenue jetty, or along the railroad tracks at the Beanery. The car edges forward in fits and starts. It *is* Barber, Elia sees as he raises his binoculars to watch. The juncos on the road in front of him are identifiable in the dim light—their white outer tail feathers flash in the headlights. The other small birds are probably goldfinches and sparrows, though it's too dark to tell. The larger birds are mostly robins, calling *chuk-chuk, chuk-chuk, weeep, chuk-chuk, chuk-chuk,* as they fly up from the road into the trees and back and forth over the road in clusters of a dozen or more.

Barber swings around to park in front of Elia's car and steps out. "Can you believe this?" he asks.

"I've never seen anything like it."

"It's the heaviest flight of the fall. Without question."

"I hit one," says Elia in a quieter voice. "A juv white-crowned. I feel terrible about it."

"I don't know how I *didn't* hit any," says Barber, lighting a cigarette. "They're all over the road."

Golden

❖

Under a three-quarter moon, they walk down the horse trail toward the dike. The air is damp, leaves are glazed with frost, and both men are wearing gloves. Barber is taller and older—in his forties, though with his size-28 waist and wiry build, he is as thin as a boy and moves with a younger man's quick and jabbing motions. He wears blue jeans, a denim jacket with a wool collar, and no hat. His long, curling hair is a mix of blond and gray. Elia is ten years younger and six inches shorter, with a thick black beard, neatly trimmed. He could pass for a museum director or a European movie producer, one who has made an effort to dress so he appeared casually attired. His CMBO visor, set high on his forehead, points at the sky. The hood of his gray sweatshirt has been folded back on his shoulders.

Elia and Barber are in the minority among Cape May's best birders in that they both have full-time, forty-hour-week jobs that have nothing to do with birds. Elia works in the back-office operation of a Philadelphia bank; Barber is a researcher for Rutgers University's Shellfish Laboratory in Port Norris, New Jersey. Each spends virtually all his free time at Cape May, however, and Elia has already this fall taken his yearly three-week vacation at the Point.

Juncos of both sexes fly up from the trail. A hermit thrush scurries into the center of the path, snatches an insect, and swallows it while the birders close to three steps. Unidentifiable sparrows scurry through the woods on both sides of the trail. Two kinglets, both showing the little diamond of red at the top of their heads, pose on a phragmites stalk.

"Listen," says Elia. "The birds in the phrags sound like rain: *swish, swish, swish.*" He lifts his binoculars to peer into the dark. "Kinglets and swamp sparrows, I think."

At the dike, the sky opens up, the light brightens, and the view is spectacular: bluejays, flickers, phoebes, cedar waxwings, kinglets, goldfinches, yellow-rumped warblers, and robins, robins, robins.

Elia scrambles to the top of the dike first. *"Unbelievable,"* he says, sweeping his binoculars through the sky.

"That thirteen-thousand-bird flight a couple of weeks ago was nothing compared to this!" says Barber. "We've got at least fifteen thousand birds in the air right now."

"At least."

Birds are flying in all directions. Two flocks of fifty goldfinches cross overhead, one group heading southeastward, from the Bayside back into Higbee's woods, the other heading northwestward out of the woods and up the Bayshore. Higher up, a loose flock of several hundred robins is heading due north. A hundred yards still higher, another flock of several hundred robins is heading southwest. Flickers and bluejays head west after leaving the woods—three flickers out, one bluejay out, two more jays out, three flickers, another jay, two more flickers—and just as quickly other flickers and jays appear from the east to land in the same lines of trees—two in, one in, four in, three in. Calling *zeeeet-zeeeet,* twenty cedar waxwings leave together from an oak tree next to the dike; a minute later, a mixed flock of kinglets and yellow-rumped warblers drops from the sky to occupy the same branches the waxwings did.

"This is the phenomenon of all time," says Elia.

"The most birds I've ever seen in my life!" says Barber.

The scattered nature of the flight on days like this is good evidence for Paul Kerlinger's theory that the Point's "morning flight" is more a dispersal of misplaced birds than the straightforward southbound migration it is usually assumed to be. Few of these birds were present yesterday, and the majority have probably arrived in the last few hours before dawn.

Nocturnal migration is the rule among most groups of birds. "The atmosphere at night is much more conducive to efficient flying," Kerlinger has explained. "Flying by day, a bird has to constantly correct for the turbulence caused by thermals. Every gust of wind— up, down, sideways—forces a correction, and every correction costs energy. At night, the laminar flow is much better, and just as important, the air is much cooler."

At rest, birds have body temperatures very close to humans, about 100 degrees Fahrenheit, but during the enormous muscular effort flight requires, their temperatures can rise as high as 115 degrees. "A body temperature like that would kill a human being," notes Kerlinger, "and it would kill a bird too if it couldn't disperse the heat quickly. In normal flight, the heat is dispersed from the capillaries just beneath the skin. During migration, though, when birds have added a layer of fat for fuel and they're flying for much longer stretches of time, it's harder for the capillaries to release the heat, so migrants need the help of *evaporative* cooling—the fat changing to water and the water cooling their bodies. You get a lot more evaporation during the night than you can during the heat of the day."

The one problem caused by nocturnal migration seems to be a kind of short-distance disorientation. Night-flying migrants can follow whatever compass direction their flight path requires in the dark (even on overcast nights without stars), but as their night of travel nears its end, they have trouble locating the precise kind of feeding and resting area they need, even on nights when the moon is full. A wood thrush might be able to fly from Long Island to Cape May Point in the dark; but once it's reached the Point, locating a stand of deciduous trees of just the density and undergrowth thrushes need is more of a problem. "That's why the owl banders catch wood thrushes in their nets out in Meadows salt marsh at four in the morning," says Kerlinger. "And that's why standing on the dike at Higbee's in the morning you can see birds flying every which way. Those birds have come down in the dark, not really sure where they are, and at first light they have to move again, to find whatever feeding area their species needs."

"With this many birds around today," says Elia to Barber on the dike, "we have *got* to find me an orange-crowned warbler."

"Your nemesis."

"I've waited too long for that bird. It's getting embarrassing."

"Well, your timing is right. Late October to mid-November is the peak. The earliest I ever had one in New Jersey was October sixth. The most I've ever seen in one day in New Jersey was five. I think it was the tenth of November."

"One of them is all I want," says Elia. "Just one damn bird."

As the overhead flight diminishes, Barber and Elia climb down the dike's mud wall and walk back south on the horse trail, heading toward the fields on the other side of Higbee's parking lot. The light on the horse trail is much better now, and Elia identifies the sparrows in front of them: "Song sparrow . . . song . . . swamp sparrow . . . white-throat . . . song . . . hermit thrush . . . song . . . song . . . white-throat . . . field sparrow . . . song . . . swamp . . . white-throat . . . Christ Almighty!"

"Damn it," says Barber, as a sparrow flies off. "Sonofabitch got away—I think I had a Lincoln's."

"Yeah," says Elia, "I had an interesting bird, but I wrote it off as a juv field sparrow."

"There's a fox sparrow," says Barber, "my first fox sparrow of the season."

"Where is it?"

"Gone, but don't worry. If there's one fox sparrow, there'll be others."

The trail is thickly covered with yellow, red, and purple leaves, which are fallen and curling into winter litter, but most trees still hold green leaves, and a few flowers are still blooming. The green, upright stalks of seaside goldenrod hold bright yellow florets; the white of the Queen Anne's lace is only faintly edged with brown; and purple pokeberries hang in thick, full clusters on both sides of the path.

"Solitary vireo over here," says Elia. "Second oak back from the little cedar tree."

"Got it," says Barber. "Nice."

"What's that behind it?"

"A maggie? . . . Hey wait, no, it's only a yellow-rump."

"What I want," says Elia, "is a damn orange-crowned."

"You'll get it, Vince. It will come to you."

Barber's most famous birding accomplishment involved a warbler, a bird he found alongside a dike road on the St. Johns River near Melbourne, Florida, on March 30, 1977.

Barber was living in Cocoa Beach at the time, working as a butcher near Cape Canaveral, birding Brevard County five or six mornings a week, and he knew at a glance he'd found something strange. It was a midsized warbler, smaller than a yellow-rumped, larger than a parula, but its plumage—grayish head and nape, yellow-ish eye ring, gray and yellow throat, olive back, olive-gray wings, and a faintly yellow forehead—was unlike any Barber had ever seen. He managed to keep it in sight for ninety minutes, photographing it at close range and studying it from various angles as it foraged with other warblers in a stand of willows and Brazilian pepper trees.

The closest resemblance was to common yellowthroat, but yellowthroats have bright yellow undersides, rounded tails, and straight, dark bills. This bird was grayer overall than even the drab-best yellowthroats, its tail was notched, and its curved-down bill showed a fleshy base. The slight wash of yellow on head and breast was just enough color to eliminate orange-crowned warbler, the drabbest of Florida's regular warblers. Eventually, Barber drove to a telephone and called two other local experts. Both refound the bird later than morning, but neither could say what it was.

Over the next few months, Barber showed the photographs to other experts, but none could tell him the bird's identity. Most were reluctant to even offer a guess. Barber himself had no idea what the bird was until one evening, a year later, when he was idly leafing through Bond's *Field Guide to the Birds of the West Indies* and a sentence on page 188 caught his eye: "The immature female is very plain, the upper-parts olive-gray, the underparts whitish with a faint yellowish wash; a pale eye-ring; faintly yellowish on forehead."

Barber read the sentence again—and again. The sentence was the last in a one-paragraph description of *Vermivora bachmanii*, Bachman's warbler, North America's rarest songbird, a bird that once wintered in Cuba and nested in widely scattered swamps in the southeastern United States. No nesting attempts have been reported since the early 1960s, no current breeding area is known, and its total world population is now almost certainly less than fifty individuals and perhaps less than ten. The bird has been seen so infrequently since the 1920s that its biology is less well understood than that of such extinct species as the passenger pigeon and the Carolina parakeet, and its female plumage is misleadingly illustrated in most field guides.

Barber began studying all the information he could find about Bachman's warblers, but he kept his suspicions about the Melbourne bird to himself, until he read a report about a sighting of a female Bachman's on the species' wintering grounds in Cuba in 1980. He mailed his photos to the authors of the report, who passed them on to Roger Pasquier at the American Museum of Natural History in New York. Pasquier wrote back to say he thought there was a strong possibility Barber's bird was a Bachman's. It was paler than the female seen in Cuba, but the eye ring and the long, decurved bill were telling features, and two old specimens of female Bachman's at the museum, one collected in December 1905 and the other in March 1902, closely resembled the bird in Barber's photos. "The latter bird is labeled 'immature,'" Pasquier noted. "I don't know whether this means anything biologically, [but it] does indicate . . . that even in March some females resemble the bird in your photograph in that they have very little yellow on the undersurface."

Still, Barber hesitated to claim an identification. It bothered him that none of his photos showed the bird in direct profile. The upper bill was definitely decurved, but was it decurved enough? In 1982 one of the two observers Barber had called to the scene in Melbourne visited him in New Jersey, and they went together to Philadelphia's Academy of Natural Sciences to look at study skins. Pulling out tray after tray, looking at specimens of Bachman's warblers and all other remotely similar species, Barber finally grew certain of his identification. The telltale clue was a subtle field mark stressed nowhere in the literature: the grayish area in front of the wings, clearly shown in the photos and featured on no other similar species. At last, in 1984, seven years after the sighting, Barber wrote a report of his find for the *Florida Field Naturalist* and claimed his identification. The journal published the two most diagnostic photographs—both blurry half-profiles.

These photographs may be the last that will ever be taken of *Vermivora bachmanii*. Since 1977, fewer than half a dozen sightings of the species have been reported in the United States or Cuba. All authorities agree the species is on the brink of extinction—or has now fallen over the edge.

When *Doonesbury* cartoonist Garry Trudeau satirized birders a few years ago in a long sequence showing Congresswoman Davenport's husband, Dick, racing to Yosemite National Park in pursuit of a Bachman's warbler, Barber took his share of heckling from his friends. Trudeau's series ended with Dick Davenport, who had crumpled on the ground in cardiac arrest, reaching for his camera's trigger as the long-lost Bachman's perched on the end of his lens. "Tweet?" asks the warbler in the final frame, as it peers down at the birder in the grass.

"Yeah, some people have said that's supposed to be me," says Barber. "But I didn't die of a heart attack when I saw the bird. And I wasn't in Yosemite when I found it. What the hell would a Bachman's warbler be doing in California? That's at least two thousand miles west of their nesting grounds."

"Don't you feel you could just raise your bins and you'd find a bird?" says Barber to Elia. "There . . . I just did it."

"I just did it, too," says Elia, "and another bird flew in front of that one. There's a swamp sparrow . . . another one . . . white-throat . . . song . . . *another* swamp. . . ."

"You notice we aren't seeing any more juncos now? They were all over the dike. The birds are spreading themselves out to feed now, finding their own habitats."

"You notice we're not seeing any white-crowns at all?" says Elia. "The only one today is the one I hit, goddamnit. I think their peak is over for the season."

"With all the white-crowns we've had this fall I've seen only two adults."

"I've had three—two together at the Meadows parking lot and another at Jeff's feet last week."

"Yeah, that's one of the two adults I saw."

As a white-throat sings from the woods, *old sam, peabody, peabody, peabody,* Barber pulls off one glove to light another cigarette. "What the hell?" he says. "Vince, look at this." His thumb has a nickel-sized stain of blue. "That's pokeberry. I'll bet a bird shit on my thumb." He is delighted. "I *know* that's what happened. Last time I lit up, one of them got me—a robin, probably. With this many birds around, the odds of getting shit on are pretty high."

A mile southeast of Higbee's, CMBO Director Paul Kerlinger parks his truck in the reeds alongside Sunset Boulevard at the entrance of the driveway that leads to the Far North Banding Station. He has come to talk again with Chris Schultz about computerizing the banding information: "Playing the prick, that's my job." Walking to the blind, he realizes it's going to be hard to concentrate on data base entry. Everywhere he looks—roads, sky, woods, trail—birds are moving.

He stops at the edge of the field beyond the station's outermost mist net and calls ahead "Clear?"

Schultz steps out of the blind at that moment, his Fu Manchu lifted in a smile. "Clear," he says, and points. A red-tailed hawk, just captured, is under the net in the far bow.

Kim Stahler is inside the blind, working the lines to the lure birds. She is a novice bander from Ohio, in her second month at the Point. While Schultz disentangles the hawk and resets the bow, Kerlinger sits down on Stahler's left, rests his elbows on the ledge beneath the blind's window, and scans the sky with his binoculars.

"This is the eighth tail we've caught already this morning," Schultz says as he steps back inside with the hawk in his hand.

"That's nice," says Kerlinger, pointing straight ahead to a huge, dark bird two hundred feet high, a quarter-mile out. "Are you going to trap this golden eagle next?"

"Flap the pigeon, Kim!" says Schultz, bending forward to look. "Flap the pigeon!"

Stahler pulls the line to the central lure bird. It wobbles upward, beating the air, then descends. Stahler tugs once more, and

Golden eagle

the pigeon wobbles upward again. The golden eagle's head never turns as the bird drops out of sight over Pond Creek.

The banders remain hopeful. Two golden eagles were seen from the hawkwatch platform yesterday, but both stayed high, apparently crossing the Bay without hesitation. This eagle is low and actively hunting, probably searching for ducks in the creek's reeds. It is almost certainly hungry.

At Higbee's, Vince Elia plucks a golfball-sized red fruit from a low-hanging tree. "You want a persimmon, Bob?"

"Not if you have to pull that hard."

"No, these are just right. They're good." He swallows one in two bites and plucks another. "Mmmmm, I love 'em."

"I don't know, Vince. There seemed to be a lot more of them there last week. Who else is eating them besides you?"

"Animals like them, too. Raccoons. I don't know what else."

"Fox sparrow," says Barber pointing.

Elia tosses the persimmon over his shoulder and swings his binoculars up quickly. "Damn, I missed it. . . . What's this? Here's an oriole, Bob."

"This is the traditional oriole spot," says Barber.

"A day like this is the kind of day Witmer Stone liked."

"Me too," says Barber. "You don't need to see anything good, just the flocks of birds flying over. How *could* you find something rare on a day like this? Your vision is totally clogged with all the regular birds."

A couple walks up. The woman holds the hand of a toddler, who is wearing her binoculars, which hang low on the too-long strap, almost touching the ground. The man carries a sleeping infant in a backpack. "Listen," he says to Elia. "Do you hear a winter wren?"

The sound from the bushes is *schuck-schuck,* like a muffled robin. "I think that's a hermit thrush," says Elia. "It's real close to a winter wren's call, but it's more of a sucky sound. The winter wren is more a *sick-sick.*"

"Here it is right here," says Barber, "eating porcelain berries." The thrush, balanced far out on a tiny limb which curls with its weight, leans over headfirst and gobbles a fruit from the cluster beneath its feet. The berries are as bright and as varied as jelly beans—peppermint, lime, grape, cherry.

"Look at this," Barber says peeling off his glove to show the man his stained thumb. "A bird shit on me. That's how many birds are here today: so many that getting shit on is something you have to expect."

At Far North Station, the eagle is still out of sight. Kerlinger picks up the radio and calls Mike Maurer in the Hidden Valley Banding Station, half a mile northeast. "Can you see this golden, Mike? It dropped down over the Creek."

"We haven't seen an eagle all morning, Paul. All we've had so far is red-tails and sharpies."

Kerlinger calls Jeff Bouton next. "Far North to Hawkwatch. You there, Jeff?" The platform is nearly a mile southwest of Pond Creek, but Bouton has a much wider horizon than any of the banders in the stations, and an eagle a mile away is well inside his identification range.

"I'm here."

"We had a golden eagle a couple of minutes ago, low over the Creek. Can you see it?"

"Negative. It must not be high enough."

At Higbee's Elia spots a phoebe without a tail. It's a first-year bird, with a lemon-yellow breast, sitting in the top of a maple tree, apparently undaunted. "He must have had a close encounter with a sharpie," says Barber.

They turn the corner and Barber points to the birds in the path. "I don't know what it is about this spot, but there's always birds on the ground here."

"Yeah, maybe seed of some kind."

"Well, there's always yellow-rumps too."

"Maybe ants?"

"I don't know what it is."

Elia sweeps through the sparrows diligently: "Swamp . . . field . . . white-throat . . . field . . . white-throat . . . swamp . . . song . . . song . . . song . . . field. . . ." He looks over at Barber. "If we have this many sparrows here, just think how many sparrows are at the Beanery right now."

Barber grins, "Yeah. Probably a better chance for an orange-crowned there, too."

Elia raises his eyebrows. "What the hell are we doing here then?"

At Far North Station, the banders spot the eagle again, to their far left this time, west of Pond Creek. Stahler pulls the cord to the lure bird at the main bow, and a red-tailed hawk swoops in from the right and grabs it. Schultz runs out to take the hawk out of the trap, and Stahler stands to go with him. Kerlinger grabs her by the rear pocket of her pants. "Stay here. The fewer people the eagle sees, the better."

Schultz carries the red-tail back into the blind, slides it into the doubled coffee can container, and sits down. Kerlinger lifts his binoculars to the golden, Stahler yanks the line again, the pigeon flaps into the air—and a Cooper's hawk swoops in from overhead and grabs it.

The blind shakes with banders' curses.

* * *

Barber and Elia park their cars side by side at the Beanery near the railroad tracks, then walk together to a circle of birders in the pumpkin patch a hundred yards off the road. Bird photographer Alan Brady holds a British magazine's photo of a strangely proportioned shorebird—plump belly, medium-long legs, and short bill. "Anyone who wants can have ten guesses. You'll never get it."

"Can I guess?" says Barber.

"You I give one guess."

"A juvenile knot."

"Nope."

"What is it?"

Brady chuckles. "A curlew sandpiper with its bill snapped off."

The birders groan and moan, except for Barber who takes the magazine for a closer look. *"Damn,* I should have gotten it," he says, tapping the picture with a forefinger. "Look: you can see the bill's broken edge."

"Hey, everybody!" says Elia, "I've got a golden eagle out here—low over Pond Creek."

At Far North, Schultz has banded and released the red-tail and the Cooper's, and he has taken over the line to the lure bird. Kerlinger, elbows propped on ledge, binoculars to eyes, is watching the eagle's head. To save the pigeon's energy they are trying to flap only when the eagle is watching. "Looking away," Kerlinger narrates, "still looking away. . . . *Now,* flap! flap! . . . Okay, looking away . . . looking away. . . . Flap! flap! . . ."

The birders at the Beanery pumpkin patch are sweeping the southwestern horizon with their binoculars. Elia's eagle is just visible above the trees.

Golden

❖

"That was a different bird than the two we saw yesterday," says Barber as it drops from sight. "Kind of intermediate in color. It wasn't as dark as the dark one yesterday, but its wing patches weren't as prominent as the light one's."

"Hey, here's a rough-leg," says Elia, pointing farther east.

The birders change direction in unison. A light phase rough-legged hawk is flapping left under the first cloud of the day. "Nice pick, Vince," someone says. "You're hot."

"Oh, yeah? How come I can't find an orange-crowned warbler then?"

The eagle is more than four hundred yards out from Far North Banding Station, and a hundred feet high, when it banks one last time, sets its wings, and dives.

"It's on you, Chris," says Kerlinger. "*It's on you.* Here it comes."

Schultz centers the lure bird in the trap and tightens his grip on the line. In nine years and many thousands of hours of banding, he has never caught an eagle, but he has heard they can hit a lure bird so hard the line is yanked from the bander's control.

At two hundred yards away, the eagle drops in a roller-coaster descent to ten feet off the ground, just clearing the tops of the hay at the edge of the field. At one hundred yards, it drops again—to a yard off the ground—and gains acceleration in the aerodynamics of ground effect. At fifty yards, hurtling forward, feathers rippling, it sweeps its black legs down and forward, then—*bam!* The straw and dirt explode as the eagle hits the lure bird and almost somersaults over it, toppling onto the ground.

The eagle backflaps to regain its balance, turns around, and steps onto the stunned pigeon.

"Pull your wings in," Schultz whispers. "Pull your wings in. *Pull your wings in.*"

[215]

The eagle ducks its head to the pigeon, pulls its wings in, and Schultz springs the bow. *Thump!* The bow whacks the ground and the eagle falls over, tangled in the mesh.

Stahler and Schultz leap into each other's arms and jump up and down, whooping and screaming. Then they charge out the door to the net. Kerlinger, still in his seat, picks up the radio. "Houston, this is Tranquility Base."

"What's that, Paul?" asks Bouton on the platform.

"This is Tranquility Base, Houston. The Eagle has landed."

For the first time all fall Al Nicholson is visiting the platform. Painting has become difficult in the frosty early morning air, so he has spent several recent mornings scanning the skies from various lookouts, the first concentrated hawkwatching he has done this year. He is standing with Clay Sutton, talking of the goshawk and the rough-leg he saw yesterday, when Bouton's radio reports the eagle capture. Bouton and others pump their fists. "Schultzie's trapped a golden!" "Wow!" "Fantastic!" Nicholson is angry. "That's a damned shame," he says, and without another word, he turns and goes, walking down the ramp to his car and driving off.

Sutton watches his former mentor leave with mixed emotions. Although he has come to disagree with many of Nicholson's radical views on conservation issues, he still shares Nicholson's distaste for the banding of hawks. Trapping a raptor is somehow a violation of the animal's spirit, Sutton believes, though he understands—better than most banders—the scientific value in the data accumulated in the process. He has never visited a blind in operation or watched a hawk be captured. He and Chris Schultz are friends, but when Schultz gives his demos, Sutton often stands at the far end of the platform with his back turned to the show. This morning, however, he climbs down from the platform to watch the demo from ground level. The only golden eagle he has ever seen close at hand was a bird he once found dead in a field in California, shot by a gunner who had cut off the bird's talons, apparently as a souvenir.

Paul Kerlinger arrives in his truck, just as the 10:00 A.M. banding demo is scheduled to begin. He has driven to the other banding stations to tell them to release all the hawks they were keeping for Schultz's demo, stopped at the CMBO office to alert Pat Sutton, and driven to the banders' house to rouse the two owl banders, Katy Duffy and Pat Matheny.

One bander is already waiting in the parking lot, with a sharp-shin on the front seat of her car for Schultz. Kerlinger pulls the sharpie from the can and holds it up for the people gathered next to the pavilion. "Folks, save your film. This is a sharp-shinned hawk. You can see it's been banded. I'm going to release it." The bird flies up and away. The people watch it go, then look at Kerlinger, confused. "If you wait a few minutes," he says, "we're going to show you a golden eagle."

Pat Sutton, Duffy, Matheny, and several hawk banders from other stations arrive, a couple of them running, and the crowd grows quickly to more than a hundred, most packing close to the picnic table Schultz will stand upon. Half a dozen people step onto the tops of the picnic tables nearby. Jeff Bouton, camera in hand, climbs onto the roof of his car.

Schultz and Stahler arrive in Schultz's truck. Stahler holds the eagle in her lap, one hand gripping each leg. When she opens the door of the truck, she cradles the bird like a baby or a football. Schultz walks through the crowd like a bodyguard escorting a royal couple. He points to a woman and her dog. "Keep that poodle on a short leash!"

"Whooo-eeee! Whooo-eeee!" Stahler cries, following Schultz and beaming, "Ain't this great?" Cameras click like Geiger

*"The ninth golden eagle in
twenty-two years of banding."*

counters. When Stahler changes her grip again to hold the eagle's legs
at her waist, its head reaches her nose. Its feet are as bright as lemon
sherbert and twice the width of her fists; the talons are as thick as her
fingers. The size-9 band is the largest the banders have, reserved for
eagles and great horned owls, and unlike all smaller bands, which are
simply squeezed together, it is held together with a pop rivet. The
old-fashioned size-9 bands had to be discarded when banders discovered that golden eagles could bend and open ordinary bands with
their bills.

The eagle's torso has an athlete's lines—rounded muscular

shoulders and a broad back above a narrow waist—and, as the bird spreads its huge wings for balance, it seems Stahler might find herself airborne if the bird flapped hard. But the bird looks self-assured, even haughty, as it folds its wings, shakes them into place, and turns its head left and right to look around at the crowd and then off into the distance. The plumage is dark, as black as a vulture's, though here and there the edge of a white feather catches the light. The hackles on the back of the head are erect and pale gold. The gold continues over the crown of the head down in a V over the dark eyes, so the bird seems to be frowning. As the hackles blow in the wind, the eagle has the look of a long-haired conductor glaring at an audience that has interrupted his concert with their noise.

It's only the ninth golden eagle captured in twenty-two years of banding at the Point, "a fitting finale for the last demo of the season," Schultz tells the crowd, and the first he has ever captured anywhere. "I've been waiting for this for a long time." Like all other goldens banded at Cape May, it is an immature male; a female would be even larger. Schultz holds up the banded leg to show the hallux, the rear toe, the most potent of the eagle's weapons. It looks like a grappling hook sharpened for a gang war.

"I'm glad I'm not a jackrabbit," says Stahler.

Schultz tells her she also might be in trouble if she were an antelope. Stories persist that eagles kill pronghorn antelopes on the western prairies. Other stories say Mongolian falconers once trained eagles to track and kill wolves. "Eagles have much sharper eyesight than humans," Schultz says.

"This eagle can see every car in that parking lot," says Kerlinger.

"And every piece of paper on every windshield," says Schultz.

Clay Sutton, standing at the rear of the crowd, raises his binoculars high, then calls out, "And if you want to see a real eagle, look up right now." A second golden, white wing patches flashing, is pumping out over the Bay in the bright sky. Only a few in the crowd bother to look upward, and even fewer seem to locate it. All quickly return to Schultz's bird, except Sutton, who follows the second eagle until it is a dot on the horizon.

As Schultz's talk ends, several banders step up onto the table to hold the eagle. Each takes it carefully, one leg at a time, oohing and aahing at the weight and the size of the feet. The eagle remains calm, turning its head slowly, eyes studying the distance. Some spectators drift off. Those taking photographs run out of film.

Schultz takes his eagle back, walks to the sidewalk beyond the crowd, and almost exactly one hour after he caught it, he releases it. The huge bird flaps awkwardly away, tail spread wide, legs dangling, wings downstroking deeply. It clears the phragmites between the platform and the pond by less than a yard, and circles left, still pumping over the platform, then continues away, low over the pines to the northeast. Those remaining whistle and applaud.

Soon, only the banders, talking in a couple of circles, remain obviously excited. The parking lot is clearing as most of the demo crowd leaves. Schultz and Kerlinger are speaking with a reporter from the Philadelphia *Inquirer*, who has come to the Point by chance today to write a story on the banding operation. Schultz tells of a time last fall when an eagle landed at his trap but stayed just far enough away from the center post so that he felt he couldn't trigger the hoop. He had to let that one go. "I've been waiting for this for a long time," he says again.

Vince Elia and Bob Barber have arrived and joined Sutton, Bouton, and the others on the platform. It has already been the best day of the fall, but all are scanning the sky diligently, thinking as birders do everywhere, no matter how good their luck has been, *What's next?*

Chapter Nine

OWLING

AN HOUR BEFORE DAWN ON A SATURDAY MORNING IN MID-NOVEMBER, Pat and Clay Sutton have a hurried, stand-up breakfast in their kitchen: orange juice, cereal, and coffee. Clay has a day free from work, Pat is not due at her office until nine, and they have planned an owl walk in the forest of cedars, oaks, and pines that stands between the platform and the South Cape May Meadows.

Pat holds one finger up to note a great horned owl woofing far off in the distance as they step outside. It's a bird they hear almost nightly from their house. "I just *love* that sound," Pat says. "The call of the wild."

They load their car—two scopes, two cameras, two ther-moses of coffee, two pairs of rubber waders—and leave under a starry sky. Their headlights find a woodcock in the road, which crouches as they approach, then explodes into flight, straight up and out of sight. "Today is the first day of small game season," says Clay, "and I almost forgot. When I was a kid, this was the most important day of the year for me, better than Christmas. Then my father died, development came to Cape May County, and hunting lost its thrill for me."

The banders are having a party tonight at the Choate House, their living quarters at the Point, and the Suttons have been invited. "I think banders get hooked for the same reasons I got hooked on hunting and fishing," says Clay. "It's the same hunter's instinct—outthinking the animal, luring it in, and capturing it. I can understand that, but I've never spent ten minutes in a blind, and I never intend to. I can't stand the thought of sitting in a little rat-trap box, trying to look up at the sky through a tiny slit. And a bird in the hand loses its magic for me, somehow. That eagle at the demo was smaller than I expected, and tamer. The one that flew over at fifteen hundred feet was a lot more exciting."

The conversation turns to Jeff Bouton, who is wearing down, the Suttons agree, as all counters have at this time of the season, except Frank Nicoletti, who seemed never to tire. Bouton has reported an hour late to the platform the last couple of days, and yesterday Paul Kerlinger chewed him out. "It's understandable, really," says Clay. "Jeff has only had two days off all year, and he gets paid next to nothing. People think he's only counting birds, which is supposed to be fun, so he ought to do it for nothing. One year I spelled the counter for three days. Three blue-sky, cloudless days. It was brutal. By the end of the second day my eyes hurt so bad it felt like I had grit under my eyelids, like I'd been walking through a sandstorm. I went home that night, turned off all the lights, and just lay there in the dark with my arm over my face. I didn't want to look at another hawk for the rest of my life. Nicoletti spoiled us—the Iron Man who never wanted to be relieved. We shouldn't expect Jeff or anyone else to stand there every day like Frank did."

The sky has grown red by the time they reach the Lighthouse.

Owling

❖

They park next to the empty platform, walk up the ramp, and scan the beach and the gray-and-black, wind-chopped ocean. A single bird flies by, a snow bunting roller-coasting southbound over the dunes. Winter is near.

Pat leads the way along the path to a wooden boardwalk and then into the cedars and pines to the east of the platform. The forest seems dead still, not a bird chip to be heard, and her stride has a happy bounce. This is the kind of birding she prefers, when you must sneak up on the birds to see anything at all.

Clay's morning owl walk lasts about a hundred steps. At the first opening in the canopy, he stops to scan the clouds, and a V of cormorants appears. "I think I might peel away here," he announces. Pat plunges onward. "Clay and I are really incompatible that way. He needs to be able to see the sky at all moments, he loves hawks so much. Owl-watching, you've got to get deep into the woods."

A flock of bluebirds passes overhead, calling plaintively *tru-*

A November morning on the hawkwatch platform

lee, tru-lee. A couple of sharp-shins flush and fly away squawking in their ratchety voices. Here and there a tree creaks. "We're walking by saw-whets and long-eared owls right now." Sutton whispers. "You have to keep telling yourself that, and you have to check every tree. The morning light coming through the branches gives them a lacy look, so you watch for something dense, something that doesn't fit—an out-of-place clump in the laciness. Long-ears, saw-whets, barn owls, screech owls—they're right here, all around us."

Owls are much more numerous than most people, even most birders, generally realize. The screech owl is probably the most common raptorial bird in North America—more numerous than any hawk, including kestrel, red-tail, and sharp-shin, all very common species. It lives wherever trees grow, from Maine to California, nesting in woodpecker holes and other natural cavities in forests, farms, orchards, suburban parks, and even in birdhouses in the backyards of city-dwellers who do not suspect its presence.

Other owls are successful over large ranges in the United States. The barn owl nests in at least forty states. The long-eared owl and the saw-whet owl nest or winter in every state except Hawaii and Florida. The short-eared owl nests or winters in every state, including Hawaii. The great horned owl, *Bubo virginianus,* has by far the widest range of distribution of all three thousand species of birds in the Western Hemisphere. It lives year-round in all provinces of Canada and all states in the continental U.S., and nests from Labrador to Tierra del Fuego, in every country in North, Central, and South America. The only vertebrate with a larger range in the New World is *Homo sapiens.*

Early taxonomists believed owls were closely related to hawks; the term "raptor" still applies to both groups, and their apparent similarity has become the standard example of convergent evolution in high school textbooks. Both feed primarily on live, large prey which they catch through speed and stealth, overpower with their strong legs, and kill with sharp talons. Both have wide, round eyes set close together and facing forward on the head, giving them good binocular vision, and both have thick, hooked bills designed for tearing meat.

Actually, however, hawks are no more like owls than destroyers are like submarines. Where owls are at their best hawks cannot go. No hawk hunts at night. The most dark-adapted of all, the bat hawk of Africa and Indonesia, feeds for only an hour or so a day, at dusk, on the first bats of the evening (and the last swiftlets and swallows), while there is still enough light to hunt by vision alone. Very few birds of any kind except owls hunt in full dark, under moon-less skies. Even the owls' closest living relatives, the *Caprimulgiformes,* the group that includes the whip-poor-will, the chuck-will's-widow, and other insect-eating nocturnal birds, depend on their vision alone to find their prey and are primarily crepuscular hunters. Owls have so little need for visual assistance that most of them are more active through the middle of the night than they are at dawn and dusk.

Every kindergartner who has ever sketched an owl has noted some of the adaptations that separate owls from hawks and other birds. Owls' eyes are bigger and closer together, their faces are flatter, their heads are wider and more two-dimensional. Owls are popular with children and cartoonists, painters, and sculptors because they look so human. Like cartoon people, owls seem to be made of small parts attached to a big head. The turned-down bill looks like a nose, and the contrast of short, stiff feathers surrounding the eyes with the longer feathers on the back of an owl's neck seems to mimic the contrast between facial skin and hair on a human head.

The eyes of a great horned owl are actually as large as a human's—though their bodies are only a fiftieth of our weight. If our eyes were as big in proportion to our bodies, they would be the size of melons. The ratio of eye to body weight in an owl is the same as the ratio of brain to body weight in humans. Owl eyes take up so much room in the skull that an owl can barely move its eyes in their sockets at all. Owls must turn their heads to follow moving objects, twisting their necks around to 270 degrees of full circle. The close-set eyes give owls even more binocular vision than hawks have and so increase their sensitivity to low light. To expose the maximum amount of retina, owl eyeballs are less round and more tubular in shape than other birds'. The lens is also more spherelike and less flexible than in other birds—or in the human eye. The widespread piece of folk wisdom that owls are blind in daytime has a kernel of truth. Owls do not see quite as well as hawks and some other birds in daylight, and they are somewhat farsighted by day or night. A human's eye can change six to ten diopters when focusing from far to near; owls have an accommodation of only two to four diopters. In zoos, at feeding time, owls often must step back from the prey offered them before they recognize it and jump forward to grab it in their talons.

Owls are aural creatures. The wide skull, the flat face, the short feathers radiating from the bill to the side of the head (the "skin" of the owl's face) have nothing to do with any humanlike qualities in owls. All are adaptations for listening. Shorn of their plumage, owl skulls look more extraterrestrial than human. The ear openings are lopsided and huge, even bigger than the eyes, and face forward, cupped by the skull like air scoops on a race car. The asymmetry of the openings and the extrawide head enhance the owl's ability to detect the source of sounds. Research suggests that a barn owl can detect differences in the timing of a sound reaching each of its two ears as short as ten-*millionths* of a second. Apparently, owls center themselves on their targets by orienting their heads (and their pursuit flight) so that the squeaks and footsteps of their prey reach both ears simultaneously and are equally loud in each ear. The mouse is centered from right to left when the sounds are simultaneous; it is centered up and down when the volume is equally loud in each lopsided ear. The short, denser feathers of the owls' face help collect and focus sounds. The line of feathers next to the ears can be raised and turned on ridges of skin to collect sound from different directions. Large flaps of skin in front of the ears are raised to deflect sound coming from *behind* the head. Like rabbits, owls hear well in all directions.

All of these adaptations enable owls to hunt at night, when no other raptors can, and each seems an essential component of the detection system. Barn owls are able to capture free-running mice in laboratory conditions of absolute darkness, snatching them at killing speed with pinpoint accuracy. When feathers are plucked from their facial disks, however, the owls miss the mice. The owl with the largest and most prominent facial disk, the great gray owl of Canada and the American Northwest, hunts by ear even in daytime. In winter it flies over snow, listening for mice and other rodents tunneling under the cover, and has been observed hovering over one spot, orienting its facial disk toward the ground, then crashing through the snow to snatch mice two feet deep.

Usually, however, none of these adaptations is easy to appreciate, except in textbooks. Owls are more secretive and better camouflaged than any other birds their size. South of Canada, they spend their days hidden in hollow trees or the thickest tangles of branches, briers, and vines. The great gray owl is one of the most seldom observed birds in North America. Even the great horned, as big and as common as a red-tailed hawk, is harder to see than all but the most secretive of sparrows.

* * *

Under a cedar tree Pat Sutton finds an owl pellet—a gray ball of fur which is the coughed-up indigestible remains of a mammal. Her eyes go up the trunk and out each branch inch by inch. Satisfied at last that no owl hides there, she breaks open the pellet like a fortune cookie, and a tiny jaw pops free, plaster white, with teeth the size of sand grains. Other bones, thinner than matchsticks, poke from the fur. "I'd guess vole," says Sutton, dropping the pellet into a zip-lock baggie and then into her pocket. She crawls under another cedar to inspect a splattering of feces on a fallen leaf. "See, it looks like candle dripping. Which means it's either a hawk or an owl." She finds a circular cluster of feathers nearby and holds one up. "Junco, I think. It was probably a Cooper's or a sharpie that got him." Owls feed more often on mammals than birds.

In an hour she finds half a dozen scattered pellets, three or four dozen droppings that might have come from owls, and one cedar which, because of the extent of the spattering on the leaves beneath, she is certain has been a roost tree in recent days. No owl sits there this morning, however, not even a promising clump. The park trail has been left far behind. Often she is on all fours, sometimes crawling on her elbows beneath cedars and pines like a Marine moving up a beach under enemy fire. The branches snag the hood of her jacket, poke down her neck, snap back and slap her in the face. At one point she finds a break in the trees and stands up: "Okay, where are we? Which direction is the sun?" Most of the time, though, she seems to be in territory she knows very well. She points to a hole where she once found a fox den and twenty steps later comes upon a fallen log which the property owner has wrapped in a rubber sleeve. "Maybe he's trying to preserve it, I don't know. I showed him a salamander under that log a few years ago, and he was fascinated."

She fingers the branch of a cedar gently, as if rubbing silk. "I once saw a long-ear on this very branch. It was great! Long-ears, when they see you, get all elongated. The body gets all stretched out, their ears stand straight up. It's wonderful to watch. There's so much tension on both sides, every second seems like an hour. Then they flush, and *poof,* they're gone. Saw-whets don't flush so easily. They're so small and hard to see that they're used to people walking by without noticing them, so they just sit there looking back at you, like little stuffed toys.

"I was crawling on the ground one time when I spotted a pellet on the ground. Then I spotted another. My hands started to shake. I lifted my head, and there was a saw-whet poised on the branch, three feet away, with a decapitated mouse in its talons.

"I once walked up on another owl so close I didn't know what I was looking at. I wasn't even sure it was a bird. It was all stretched out and its eyes were slits. I got shaky and excited, dropped to my knees to steady my binoculars, and it opened its eyes to look at me. That's when I realized, 'Cripes, it's a barn owl!' I watched it for ten minutes, then I crept backward out of there as quietly as I could."

Other things catch her eye: a pin oak still holding its leaves while other oaks are bare; one last red persimmon in a tree holding clusters of brown, deflated fruit; the egg case of a praying mantis. "I once saw one of these eggs open in spring. The young came down a silken thread, eating each other."

It is small wonders like these that Sutton tries most often to show the groups she leads around the Point. Most are inexperienced birders and novice naturalists, who often tell her she's shown them something they haven't seen since they were children. *"What is this world so full of care,"* she likes to recite to them, *"if we have not time to stop and stare?"* The philosophy is one Sutton seems to struggle to live by herself. She regularly works six days a week, and during the most hectic times, when her New Jersey Audubon job and off-hours

conservation work add together, she will sleep only four hours a
night. One reason owls have become her favorite birds is that they
can be chased at either end of the workday.

Sutton emerges from the woods after ninety minutes, brush-
ing the twigs from her hair and pants legs, plucking the burrs from
her socks. She looks at her watch, then points to a property across
the street with a For Sale sign on the front lawn. "Look at the woods
behind that house—cedars, lichens, oaks tangled with vines. That's
one last remnant of Old Cape May; there are owls roosting in there
right now, I'm sure; and it's doomed. The Nature Conservancy made
a bid, but the owner wasn't interested. He's retiring and moving to
Florida, so he doesn't care what the next owner does with it; he just
wants to make one last bundle."

She breaks into a jog heading down the street toward the
platform, until a kettle of red-tailed hawks appears in the bright blue
sky overhead. "Goodness, look at this," she says, stopping for a
moment. "Twenty, thirty, forty, fifty. . . . I count fifty-six. Look at
how the red glows! They light up the sky." She resumes her jogging.
"That's why Clay and everyone else likes hawks better than owls.
You can *see* them."

Clay is standing on the platform shoulder to shoulder with
Jeff Bouton and Al Nicholson. The kettle is circling and building
over Lighthouse Avenue, a curling line of buteos, all of them adult
red-tails. "Seventy-five birds!" Clay tells Pat. "It's the biggest red-tail
kettle I've ever seen."

"I've seen one or two bigger kettles than this," says
Nicholson.

"I remember a kettle of sixty once," says Sutton.

"I have to go to work, Clay," says Pat, looking at her
watch again.

"Oh, stick around for another half-hour."

Owling

❖

"Bye."

While the kettle drifts westward, Nicholson turns away and scans the northeastern horizon. "No vultures coming. Unless vultures are coming from all over the sky, you're not going to get a really big flight." The count for all hawks stands at 41,892, a total usually reached by mid-October. Unless the last few weeks bring a freakish series of cold fronts, this season's total will be the lowest in the history of the count. "Cape May won't even be the leading count in the East this year," says Sutton, "for the first time ever. Montclair is over forty-nine thousand now—forty-two thousand broad-wings. They're shooting for fifty thousand, and they might make it. They're going to beat us, in any case. No way we're going to make up that many hawks in the last few weeks."

The three observers agree November flights at the Point are "finicky." "They start later, build slower, and end sooner," says Bouton. "Even on a good day they die at three o'clock like magic."

"All the factors have to be just right this time of year," says Sutton, "for any kind of flight at all. There's less lift in November, fewer thermals."

"In November," says Nicholson, the sky painter, "the raptors come on a mercurial light."

The kettle drifts northward until it is directly over the smokestack at the north end of Lighthouse Avenue, a mile from the platform, and Sutton leaves the platform to follow in his car. The smokestack is part of the abandoned magnesite plant, built to extract magnesium from seawater for firebricks. "When I first started hawk-watching it was still in operation. You always knew which way the wind was blowing—just look up the smokestack. It was a giant wind sock." The soot blackened the leaves, killed the trees, and the area all around the plant is now a wasteland of sand, low brush, and cyclone fencing. "It's ugly as sin, but it's better than condos," says Sutton.

"This is one of the last, big, unprotected open areas left south of the Canal."

He finds a trail of otter tracks in the sand: two cat-sized front paws, two larger rear paws, and the curl of the tail touching the ground alongside the paw prints every few steps. The animal apparently crossed Sunset Boulevard, the most heavily trafficked road in the Point, heading in the direction of Pond Creek, north of the plant. Several pairs of otters still hang on in the few remaining wetlands around the Point. Their silvery, fish-scale scat lines the paths through the Cape May Meadows; they are regularly seen in Pond Creek and Bunker Pond; and some birders claim to have spotted them in the waves off the beach, chasing fish in the ocean.

Sutton kicks an object free with his toe, picks it up, blows off the sand, and brushes it clean with his pinky. It is the cap of a shotgun shell. "Ten gauge, and made of brass so it will last forever. They haven't used ten gauge in fifty or sixty years—which is when they were shooting hawks right here on this spot."

Sunset Boulevard, "The Concrete Highway," was a well-known gunning area in the 1920s and '30s, and sportsmen drove from as far away as Philadelphia to park on its shoulder and fire at any hawklike bird that attempted to cross the road. New Jersey law dating back to the early 1900s protected all hawks except the sharp-shin and the Cooper's, but the law was difficult to enforce. Few of the gunners were knowledgeable enough to identify the hawks before they fired, and many were poor shots—Robert Allen, sent by the Audubon Society to study the shooting, watched one gunner fire sixty-two times to kill thirteen hawks one day in 1932. The birds were identified after they fell. The sharpies and Cooper's were loaded into peach baskets; the rest, hawks of all kinds, and owls, whip-poor-wills, nighthawks, and kingfishers, were left where they landed, many of them only crippled. It must have been an ugly scene, and the numbers are grim—1,400 hawks shot in one day in 1920; 140 hawks shot by a single gunner in one day in 1921; 180 hawks shot in one day by a single gunner (using eight boxes of shells) in 1933; 1,008 of the 8,206 sharp-shins seen passing in the fall of 1935 shot. Like many of the Point's local conservationists, however, Sutton looks back on those days almost wistfully—like a man with cancer remembering when he once faced a rattlesnake. The enemy was so identifiable in the '20s and '30s, and the problem so simple: stop the gunning and save the hawks. Today, the threats to hawk populations are far more numerous, and the result of the work of educated people who mean no harm to birds and never see the effect of their actions: pesticide poisoning, industrialization, real estate development, forest fragmentation. Sutton would like to see the area preserved as a park or sanctuary in commemoration of the hawks that died there. "I've heard Cape May County has been looking to buy the property, but the disposal and reclamation costs would be in the millions. A company in Texas owns the property. I'd love to see them donate the land. Are they just going to leave the plant here, abandoned? To be an eyesore for the next fifty years?"

The air is warming. A monarch butterfly appears, flapping and tilting low over the sand; a green darner follows, hovering ten feet overhead, then zipping away. Higher up, the only break in the blue is the vapor trail of a passenger jet. Not a single hawk is in sight.

What happened to the red-tails? Sutton shrugs. "Some of them crossed, some went north, and that was it." He claps the lens cover back on his binoculars. "That's the trouble with big flights. They don't last."

Nine o'clock on a Saturday night, two months into the Point's banding season, and the fifteen banders crowded in the Choate House kitchen have a common look: bleary red eyes, sunburned faces, peeling noses, scraped fingers, punctured wrists. Most of them also have a beer in their hands. Champagne and a cake await the return of owl banders Katy Duffy and her husband Pat Matheny, who last night caught and banded the thousandth owl of Duffy's eight years at the Point.

Tacked to one wall, beneath a strip of peeling pink paint, is a chart of the daily totals of hawks banded this year. All the numbers are double digits: 32, 40, 15, 27, 52. . . . On the opposite wall is a list of the nightly totals for owls banded: 0, 0, 6, 1, 0, 0, 3, 0 . . .

A cake as wide as a desk blotter sits on the table. An owl drawn with red and green icing perches over a script that reads, "Congratulations, Katy & Pat. 1000th Owl. 8 Years. 2.5 Million Hours."

The two owl banders enter to applause and the flashes of cameras. Matheny, a compact, muscular man with curly blond hair, steps out of his boots and walks into the kitchen in his socks. Duffy follows, peeling off her headlamp. As tall as her husband, her black hair flecked with gray, she has a graceful and commanding presence, and she waves off the applause with one long-fingered hand. Someone hands her a champagne glass, and she bends to read the inscription on the cake. "It *feels* like two and a half million hours, I'll tell you that. All those nights, all those miles, all that lost sleep."

Owl banding began at the Point in 1972, but for most of the early years it was a sideline enterprise, conducted on occasional nights by exhausted hawk banders who had been in the blinds all day and generally closed down at midnight. The first three years of all-night banding in 1977, 1978, and 1979 proved, among other things, that barn owls were regular migrants through New Jersey. Duffy took over directorship of the operation in 1980 and has been back every fall since, though she now lives in Wyoming. She works as a seasonal naturalist at Grand Teton National Park, where she met Matheny. He joined her for the first time in 1985, and they were married at the Point at the end of the season. As the park season ends in Wyoming at the end of September, they race twenty-two hundred miles across the country, trying to beat the barn owls to the Point.

"We knew as soon as we stepped outside last night," Matheny tells the group, "conditions were perfect. The front had come through the day before; the winds were still northwest and real light. Katy was at nine hundred ninety-two, but we didn't talk about it; she didn't want to jinx herself. The first check we had *five* owls— three long-ears and two saw-whets. The next check we had three more, all saw-whets, and Katy forgot about her total. I held up the last one. 'Here it is.' 'What?' 'Number One Thousand.' "

More applause. "Yeah, Katy. *Congratulations!*" Duffy waves it off again. "That was last night. Tonight there's *nothing* out there."

In the hallway leading to the living room one female bander is showing another the scars and scratches on her hands. "This one a Cooper's did yesterday, and here's where a red-tail got me last week. This one here is only from a sharpie, so I don't even remember when it happened."

Other conversations swirl, the banders tossing off stories like fishermen in a dockside bar.

"He took a 7A band."

"Is that right? I had a female sharpie today so big she wouldn't fit in the Pringles can."

"One time I was sitting in the blind at Hidden Valley, nothing much coming. I'm leaning back, wondering if I'm even going to *see* a hawk, and a sharp-shinned flew right *through* the blind—in the left door, out the right."

"Did you grab for him?"

"I was too shocked."

"One time I was taking a kestrel out of the bow at North,

and a white-throated sparrow flew right through my legs. I turned around to see what was chasing him, and there was a sharpie flying right at me. He put on the brakes about six feet from me, and hung there. We were eyeball to eyeball, him asking, 'Are you going to let me have that white-throat?' and me saying, 'No, I'm not.' Then he took off."

"Did you ever hear that story about the bander in Scotland? He was sitting on a fallen tree when a grouse came out the woods and ran in to hide right underneath his leg. The guy looked up and there was a golden eagle circling over him. He looked down to check the grouse, moved his leg about two inches, and the eagle nailed him right on the thigh!"

"*Ouch!* Did the guy ever walk again?"

Katy Duffy, eating her cake, stops and holds her palm to her nose. "My hands still smell like long-ears, from last night," she tells Pat Sutton.

"What smell is that?"

Duffy holds her hands up for Sutton, who sniffs and shrugs, puzzled. "I don't know how to describe it," says Duffy. "It's just long-ear smell."

Patti Hodgetts, another owl bander, leans over, takes Duffy's hand in hers, and sniffs. "Long-ear, all right. You can't wash it off. Red-tails leave a smell too."

"Someone should write *A Guide to Raptor Identification by Odor,*" says Duffy. "It would have to be a bander. We're the only ones who know."

"How about their breath?" asks Hodgetts. "Barn owl breath—*yuk.*"

"How about harrier breath?"

"Awful."

"*Peregrine* breath!"

Owling

❖

"Oh, *God,*" Hodgetts says, laughing. "The worst!"

A picture show begins in the living room. Joey Mason shows a video of a red-tail capture, and Clay Sutton, who has never seen a hawk trapped by a bow net, asks for a replay. Several people have slides of the golden eagle demo, and Mike Maurer, the most veteran of all banders present, has pictures of a golden eagle captured eight years earlier. "Those were the days when we believed we should never close down a banding station. If you didn't trap the bird yourself or have a partner to watch the nets, you didn't see it."

Chris Schultz inserts a tray of slides of hawks he has banded. Most of the photographs were taken in extreme close-up, at a range banders know best. A single brown eye of a raptor fills the screen. "Red-tail!" someone calls out. A taloned foot: "Peregrine!" When a sequence of tails-only photos begins, the competition in the room grows intense. As each slide clicks into place, five or six voices call at once: "Sharpie!" "Merlin!" "Cooper's!" "Harrier!" Schultz's last slide is a close-up of an aberrant kestrel tail—all red and black as in a normal adult, except for one fully blue feather in dead center. The room is suddenly silent. "Jesus, that's a primary!" shouts one bander, holding his third or fourth beer. "A wing feather in the tail!"

In the kitchen Matheny and Duffy pull their boots back on, then push out the door and down the steps. Duffy studies the tops of the trees swaying overhead. "East winds, still," she says, tonelessly. "Owls evaporate on east winds." But their nets are already open, so they must go.

The first stop is the Meadows, where they have a series of nets near the East Banding Station. The truck is still rolling when Duffy jumps out. By the time Matheny parks and walks around the truck, she has leaped the barbed wire fence and is far across the field. Only the light of her headlamp is visible, bobbing against the sky. "Nights like this," Matheny says, "are what owl banding is all about. A lucky night owling is when we catch one bird. A really good night is three birds. The kind of night we had last night happens once a year."

When Matheny catches up with her, Duffy is untangling one panel in a net that has wrapped around itself. The next panel is puffed with the wind. "It used to be," she says, "I knew less and could hope more. Ignorance was bliss."

Owls seem to move through the Point in highest numbers following northwest cold fronts, just as the hawks do. But hawk banders capture forty or fifty hawks for every owl Duffy and Matheny capture. Because they cannot use the bow traps and rely primarily on mist nets, owl banders need more help from the weather. They have most success on the first calm nights after the front. A very light wind, one to three miles per hour, is best. Any wind stronger than fifteen miles per hour or so renders the nets ineffective, and the capture rate actually seems to drop as soon as the wind reaches ten miles per hour. Duffy suspects the owls can hear the mesh whistle at that speed. They also seem able to see and avoid the black mesh of the nets even when the moon is only half full.

In eight and a half seasons, Duffy has compiled 130,000 "net hours" and captured 1,000 owls, one owl for every 130 net hours. If she could keep thirty nets open for nine hours every night, she'd average two owls a night, but to protect the nets from wear and tear, she must close them down under strong winds and leave them closed when it is raining or snowing. She also loses the last hour before dawn even on the best nights because she does not want to capture an owl so close to sunrise that it doesn't have time to find a safe roosting spot after release. Despite their talons, owls are as susceptible as other birds to predation by larger raptors. Saw-whets are tiny owls, smaller than robins; long-ears are only the size of kingfishers. Both are preyed upon by other raptors, the Cooper's hawk especially. Even barn owls, which are only as big as laughing gulls, are preyed upon by great horned owls. "We close down as soon as the sky starts to gray," says Duffy. "We don't want to trap some owl, put a band on it, and then lose it to a Cooper's or a red-tail because it hasn't had time to find a place to hide for the day."

Duffy has used other trapping methods occasionally: verbails, which are spring-released devices on poles triggered by the owl's weight, and tape recordings of starlings, used to lure the owls closer. She is also one of the very few banders anywhere to have captured a great horned owl, North America's most ferocious raptor, with her bare hands. Inspecting her nets before dawn one morning, she spotted a catbird tangled in one panel and a cluster of blue jay feathers on the ground underneath the next panel. She knelt to examine the feathers in her headlamp, wondering what predator would have gone after a feisty jay, when a great horned pounced on the catbird. It was coming back for a second meal. It bounced off the net, dropped to the ground, and jumped at the catbird again. When Duffy stepped forward, the owl did not retreat. Instead, it threw out its wings and fluffed its feathers, trying to bluff her back. Duffy waggled her left hand in the air as a distraction, dropped low, and swiped with her right hand. She knew she had a capture when the owl jammed a talon deep into her forefinger. "I was still learning then, but I didn't let go. That would have gone against every bander's instinct. I just hung on and yelled, until my partner Carol heard me and came running and pulled the talon out. *Then* we put a band on that owl."

Duffy prefers to work without gloves. "Gloves limit your dexterity. My main concern is always the bird. You have to hold firmly but gently, so you won't hurt them. That's easier to do without gloves on. Eventually you're going to get bitten and footed. Everybody does. But like the hawk banders say, 'If you can't take the claws, stay out of the blind.'" Long-ears, which tend to twist themselves into the nets in knotty snarls, foot and bite her regularly as she tries to untangle them. "Oh, it doesn't hurt that much."

Matheny snorts. "Whenever we trap a saw-whet and a long-ear at the same time, I always let Katy take the long-ear. When a saw-whet grabs you, it's like a love bite."

Their banding procedure is the same as the hawk banders'. They bring their captures to the nearest station; band, weigh them, and measure their talons and bills just as the hawk banders do; examine the feathers for age and wear; record all data; and then step outside the blind to release them. Duffy lets saw-whets go from her hand. Barn owls and long-ears, she has discovered, prefer to be released from the ground. Lifted into a tree or released from the arm, they will drop to the ground, so now she bends to place the owl on the ground and steps back. The owl whirls off in silence. One moment there is a creature of the night in her headlamp; the next there is not.

The three migrant species—barn, saw-whet, and long-eared—make up more than ninety-five percent of the owls captured. Barn owls migrate first, peaking shortly after the hawks, in early October, and they are caught most frequently in open areas in the nets with four-inch mesh, which Duffy erects in the Meadows. Saw-whets and long-ears come later, from mid-October to mid-November, when the hawk migration is winding down. The tiny saw-whet tends to stay in the woods and is caught most often in nets with smaller, two-inch mesh. The two resident species, the great horned and the screech owl, generally avoid Duffy's nets, although they are both common nesters in the area. Even the larger mesh is too small to catch great horns regularly: they trampoline out of it, their wings and feet too large to be snagged. How the screech owl, which is midsized between saw-whet and long-ear, avoids the nets is a puzzle.

Returns on banded owls are even more precious than returns on banded hawks. Duffy does well only with barn owls, probably because this is the most man-adapted species and most likely to be captured in accessible areas. Her return rate for that species is about seven percent, and she has twice recaptured young barn owls banded at their nests by Hanna and Artie Richards, two banders who live in Brooklyn. When the banders began trapping owls in the late '70s, whether barn owls migrated or not was still an open question. Some biologists believed barn owl movements were simply post-breeding dispersals: young moving away from their nest to new territories. The banders' years of counts for barns moving through the Point, and Duffy's retrapping of the Richardses' owls, have demonstrated that *Tyto alba* is at least a partial migrant.

The most mysterious of the Point's owls is the long-eared, one of the most secretive of all North American birds. It is warier than the saw-whet and, unlike the barn owl, it does not call in flight. Long-ears are rarely detected by birders anywhere. On October 29, 1987, Duffy caught one that her partner Patti Hodgetts had banded at the Point on October 27, 1985, two years earlier almost to the day. It was the first banded long-eared owl *ever* recovered for the Cape May Banding Project, living or dead. Where the long-ears that migrate through Cape May nest and where they winter remain unknown. In fact, most texts refer to the species as "mainly sedentary," though Duffy and her predecessors have been trapping long-ears passing through the Point for fifteen years.

In the fall of 1982 a young birder named Bob Russell conducted a month-long owl census from the hawkwatch platform. Using a night-vision scope, which enhanced available light a thousandfold, Russell scanned the skies around him as the hawkwatchers do by day. His most exciting sighting was a loose flock of eighteen long-eared owls departing the Point at dusk and flying out over the water. This was the first visual confirmation that long-ears cross Delaware Bay and one of the few records anywhere of owls' flying together. Russell also confirmed that Duffy captures only a small percentage of the owls migrating at the Point. He was severely limited by his instrument, which had a narrow field of view and provided no magnification, but still he counted 200 migrating owls on his twenty-odd nights of watching, from October 12 to November 10, four times as many as Duffy caught during the period. The numbers of owls migrating through the Point is likely more than ten or twenty times as many as Duffy catches, perhaps more than forty or fifty times as many. No one has attempted to repeat Russell's study

Saw-whet owl

in the years since, however. It was tough, dull work, hard on the eyes, and it remains one of the few attempts anywhere to study owl migration flight directly.

Meanwhile, Duffy perseveres with her banding project. What has she learned after eight years? "Everything matters: the strength of the wind, the direction of the wind, which size mesh we use, how far the phragmites have encroached in the Meadows, how well the meadow voles are doing, what time the moon goes down, how many cold fronts we get. Our worst year, nineteen eighty-four, we caught twenty-nine owls all season. The marshes were flooded so the voles didn't do well, and we didn't get a single good front. The next year Hurricane Gloria came through, knocked down the dunes, flooded the marshes, and we had our second lowest count ever, forty-nine owls." The weather variables mean population trends cannot be extrapolated from Duffy's counts, but still she worries. She has never duplicated the total of 214 owls she caught in her first season, and her barn owl numbers have shown a sharp decline. Over her first four years she averaged thirty-five barn owls captured each season. Over the last four she has averaged seven. Like Pat Sutton and all the Point's veteran birders, Duffy agonizes about habitat loss at the Point. "Every fall I come back and there are more people, more cars, more houses, and more lots cleared for more houses. Each parcel you lose means more fragmentation—fewer voles and mice, fewer roosting trees. Every year it seems to get tougher for owls to find what they need here."

Tonight the only moment of suspense comes after Duffy and Matheny have found every net in the Meadows empty and are returning to the truck. Duffy stops her hard march long enough for Matheny to catch up, and asks, "What was that?" Matheny shrugs. Duffy puts her mouth to the back of her hand and makes a squeaking sound. They wait. Trees bend and rub in the wind. Duffy squeaks again. Nothing. Duffy turns and goes, at a pace a jogger would admire, out of the woods, over the barbed wire, and back to the truck.

They drive to their lanes near the North Station and to the lanes at Far North and find the wind blowing so strongly that Duffy decides to close up all nets. They roll up each panel, tie it shut, return to North to tie the nets shut there, then drive back to the Meadows to close the nets there. Finished, they do not return to their bedroom at the Choate House. They park their Nissan in the driveway at CMBO headquarters and lean back in their seats to nap. "The truck is uncomfortable, which is good," says Duffy. "That makes it easier to get up. Who wants to get out of a nice warm bed to go walk around in the cold and the dark looking into empty nets?"

It is midnight. They will wake again in an hour or so. Should the wind change, they will go back to their lanes, open all nets, sleep another hour, then walk the lanes again. If they are lucky, they will catch an owl.

Chapter Ten

❖

SEASON'S END

AT 9:00 A.M. ON THE DAY AFTER THANKSGIVING THE PLATFORM IS EMPTY except for a black cat with a white chin, which crouches, tail twitching, on the corner bench where Jeff Bouton usually sits. Another black cat creeps through the weed patch next to the platform, and a tawny-colored cat has leapt to the lip of the trash can under the pavilion and dropped inside to scavenge. The appearance of feral cats out in the open is an annual late November event at the Point and as sure a sign that autumn is ending as the first snowfall. The songbirds they have preyed upon through summer and fall are mostly gone now, and the cats have become bony and desperate. Bouton has already found one dead and frozen to the ground in front of the platform.

The wind blows hard from the north this morning; the water is brown under a choppy surf. The air has turned cold in recent days, with hard frosts on several mornings, and the last warm-weather lingerers—laughing gulls, Forster's terns, long-billed dowitchers, tree swallows—have departed. The sun has also moved south. In August it rose directly behind the Meadows, coming up over Cape May City. Now it is out over the ocean, rising above the bunker and tracing a lower path through the sky.

The few birds in sight are all common winter residents. Two herring gulls stand in one corner of the empty parking lot. A chickadee hangs on a phragmites stalk in front of the platform. Half a dozen house finches stand on the pavilion roof chirping at the cat in the trash can and waiting the chance to descend to the crumbs blowing across the pavilion's concrete floor. The waterfowl on Bunker Pond are common species: mute swan, black duck, mallard, wigeon, pintail. A little bufflehead duck whirls in off the ocean, lands with a splash, and a moment later, three swans rise up, scamper across the surface on their doll-sized legs and flap into the air with stiff beats. They are headed for Lily Lake, where they will be fed white bread and popcorn. In the quiet of the parking lot, the *thump, thump, thump* of their wings seems as loud as a man beating a rug.

Bouton's flat-topped oak and round-topped oak are bare now, though some color remains in the woods on the platform's skyline. Many shorter oaks still hold curling red leaves, and several maples are nearly full, their bright yellow leaves as flat and wide as cartoon hands. The cedars and pines have become prominent, the only natural green left in view. The phragmites around the platform and Bunker Pond are now silver and mauve, their heads drooping on broken stalks.

A sharp-shinned appears left of the radio tower flying southwestward toward the Bay, and it is mobbed by at least a hundred starlings, the gray undersides of their wings flashing in unison. The starlings swing away as the hawk drops below the tree line, and join a still larger flock, which circles, then descends to land around a puddle in the parking lot's center. Two or three hundred starlings shoulder each other around the water, squeaking, whistling, and poking at the asphalt. A few wade into the puddle, shaking their breasts and spraying droplets. Then in an instant the flock goes, rising and swirling away like a dust devil.

A man drives up to the platform in a station wagon with Connecticut plates and birding stickers on the rear window. He steps out of the car, rests his elbows on the roof, studies the finches on the pavilion through his binoculars, and then turns to the totals board.

He is still reading the figures when Bouton pulls up alongside in his car.

"You're late."

"My contract ended yesterday," says Bouton. "I turned in my clickers. This is overtime now." He is unshaven and looks exhausted; his eyes are puffy and bloodshot. His ski cap is pulled down over his ears and twisted to the side. The Professional Brotherhood of Hawk-watchers patch stitched to the front—Pete Dunne's peregrine, mana-cled binoculars, and *veni, vidi, computavi*—rides over his left temple.

"The last time I was here," says the man, "you had a wheatear right over there."

"September eleventh," says Bouton, his face lighting up. "Wasn't that great? Every little gap in the pines had three or four people standing behind it. How many scopes do you think we had on that bird?"

"A couple of dozen, at least. I watched that bird myself all afternoon. I figured I'd probably never see another one."

"I'll never forget," says Bouton, "the looks we gave the guy who found it. He came running up here all out of breath. 'How often do you see wheatears here?' he asked me. I didn't know what to think. We all just stared at the poor guy. 'Well, there's one right over there right now,' he said. There was something in the way he said it. This whole platform cleared in about two seconds. And then the bird stuck around until dark, feeding in the grass just like a robin." Bouton

Cape May Point Hawkwatch

SPECIES	YESTERDAY'S TOTAL	TOTAL TO DATE	PEAK FLIGHT AND DATE
TURKEY VULTURE	—	841	94 10/20
GOSHAWK	—	13	2 10/30
COOPER'S HAWK	4	2,507	169 10/6
SHARP-SHINNED HAWK	15	20,894	1,891 10/6
RED-TAILED HAWK	31	2,275	247 10/30
RED-SHOULDERED HAWK	6	449	54 11/22
BROAD-WINGED HAWK	—	1,485	188 10/13
ROUGH-LEGGED HAWK	—	4	—
SWAINSON'S HAWK	—	3	—
GOLDEN EAGLE	—	19	3 10/29
BALD EAGLE	—	36	3 9/22
NORTHERN HARRIER	5	2,533	123 10/30
OSPREY	—	2,822	283 9/26
PEREGRINE FALCON	—	339	51 9/28
MERLIN	—	1,965	190 9/26
AMERICAN KESTREL	—	7,337	833 9/21
BLACK VULTURE	—	13	6 10/12
TOTAL	61	43,534	

wobbles his hand to illustrate the wheatear's walk, for at least the thousandth time this fall.

They climb the steps to the platform, scan the hawk-less horizon with their binoculars, and the man asks, "You going to make it to fifty thousand this year?"

Bouton leans backward from the shoulders, glaring at him. "What are we at now? Forty-three five? We'll be lucky if we make it to forty-three six. This migration is *done*. It will be the lowest total count ever. I'm resigned to that."

Only three of CMBO's previous counts have failed to break 50,000 for the season, and no previous count has totaled less than 47,000. This year's drop is due entirely to low numbers of the two most common species—sharp-shinned hawk and kestrel. Bouton has seen 20,000 fewer sharp-shins than the average for the previous twelve years and 7,000 fewer kestrels. Most other species have been above average, and since the doldrums of unseasonably hot weather in late September and early October, the daily flights have been regular and heavy. The counts for merlin and red-tailed hawk are both nearly 500 above the averages for the previous twelve years. The counts for harrier, osprey, and Cooper's hawk are all nearly 1,000 above average. The peregrine count is lower than in the previous two years, but still higher than the twelve-year average.

"People will look at the bottom line," says Bouton shrugging, "and say what a bad year we had. Montclair has forty-nine thousand, and they're going to beat us for the first time ever. They had that one big flight of seventeen thousand broad-wings in September, and we never caught up. But in my opinion it's been a good year for raptors at Cape May. We set two new all-time records. The old best-ever for golden eagles was sixteen; the best-ever for black vultures was five. The Cooper's count is the second or third best ever. The harrier count is the second or third best ever."

But the high counts of the other species seem to accentuate

the low counts for sharp-shinned and kestrel. If other hawks had a good migration this year at the Point, how could it be that Cape May's two most abundant hawks did not—unless their populations are truly down?

"Pete Dunne's theory," says Bouton, "is it's the songbirds' drop that's creating the sharpie drop. Songbirds are harder to find, and so the sharpies are crashing—and maybe he's right. Kestrels will feed on songbirds too, so that might be a factor there also. But after two years of doing this count, I keep thinking what happens at the Point is all about weather. To me, Clay's theory about the timing of the fronts makes a lot of sense. We didn't get the fronts during the peak of the sharpie and kestrel movement, so they went south on other routes. We didn't see them here because the weather didn't drive them here." He scans from right to left. A black-backed gull is the only dot in the blue. "Maybe both those theories are correct," he says. "The sharpie population is down, kestrels had a bad year, *and* the timing of the fronts was wrong."

Bob Barber pulls up in his car. He was here yesterday, on Thanksgiving, and has been one of the few platform regulars sharing Bouton's watch in recent days. His mud flaps bear a thick crust; his windshield and his binoculars are covered with dust. He has been birding since dawn. "November is my favorite birding month. I don't know why. Especially today. There is *nothing* around. It's incredible: three days of north and northwest winds and no raptors. Land birding is dead too. Nothing at Higbee's this morning, or Pond Creek, or the Beanery. Gannets off the Second Avenue jetty are about the only birds I've seen all day. But there's nothing to go home for, either. Not even a football game on TV today."

"Shouldn't red-tails be coming through still?" asks the man from Connecticut. "Are they past us already? Or are they still to the north of us?"

Season's End

❖

"Both," says Barber. "What we need is a heavy snowstorm in central New York. That would drive some more tails and red-shoulders down here."

"What I need," says Bouton, "is one more golden. That would give me twenty. A nice round number."

"I think you need six more," says Barber. "Make it twenty-five. Then you'd have a record that would stand for a while."

"I'd have to stay here till March."

"Ah, you're young," says the man from Connecticut. "Enjoy it while you can. What else can you do?"

"Go back to school. I've come just about as far as I can without a degree. It's frustrating. Some places won't even let you band unless you've got a B.S. Which is ridiculous. I *guarantee* you I know more about banding and hawk I.D. than most guys with a B.S."

Banders often joke that the only birders who prefer birding to banding are those who have never triggered a trap. Once a hawk-watcher has held a hawk in hand, banders say, watching hawks from the platform seems deadly dull. Bouton has been the sole local exception to this rule—a hawkwatcher who was a bander first—but next year he will be returning to the blinds again. He learned banding in high school, at a station outside Rochester, New York, near Braddock Bay. Last winter he helped a Braddock Bay researcher trapping saw-whet owls in mist nets on the Lake Ontario shore. This week he has applied to the Colorado Division of Wildlife in the hope of joining Chris Schultz next spring on the peregrine project, banding nestlings on their cliff-side aeries.

"I'm looking forward to coming back here next year, too. The best thing will be sitting the blinds and not having to feel responsible for every dead day. What a relief that will be. No more 'Hey, you're the counter, right? Where the hell are all the hawks?'

Next fall I'm going to stay out of sight and enjoy the flight again."

Bouton's radio crackles. It is Paul Kerlinger calling from CMBO headquarters. "Hawkwatch, we just got a report you'll want to hear about. Patti Hodgetts telephoned from the Parkway. She saw a golden eagle over Milepost Ten, and it was flying this way."

"Hey!" says Bouton, taking two giant steps toward the front of the platform. "That's my number-twenty bird!"

The next hour passes slowly. Two sharpies and a red-tail in sixty minutes. Finally, Barber and the man from Connecticut walk over the dunes to watch for gannets and scoters from the beach. Bouton goes to his car, scoops up a handful of birdseed from the fifty-pound bag he keeps in the back seat, and spreads the seeds under the pavilion. The house finches descend as he trudges back up the platform steps. He sits down, props his arms on the railing, and aims his binoculars at the scrabbling little birds. "Yeah, I think I'll stay here until the golden comes, then wait for one more harrier, check that off, walk over to my car with my head down, and drive home to Rochester. I began the year with a harrier, so I want to end it with a harrier."

The tawny-colored cat creeps out from beneath a bush, and the finches fly up and away, chirping. Bouton sits back, raises his binoculars to the horizon, and scans from left to right: wooden tower, flat-topped oak, round-topped oak, round-topped pine, merlin sticks, water tower, radio towers, dunes, bunker. Not a hawk is in sight. "We have to clear out of the banders' house this weekend. Maybe if the golden doesn't show today, I'll sleep in my car and come back tomorrow." He scans from right to left: bunker, dunes, water tower, radio towers, merlin sticks, round-topped pine, round-topped oak, flat-topped oak, wooden tower. "This happened last year too. I've been saying for weeks I couldn't wait for the season to end. Now that it's over, I'm saying 'I don't want it to end yet.' "

Season's End

❖

* * *

On East Lake Drive, Paul Kerlinger carries boxes of sweatshirts from his truck up the steps of CMBO's headquarters and stacks them in the vestibule, which is already crowded with bird books, posters, brochures, and boxes of visors and T-shirts. Kerlinger has been running errands all morning—mailing off certificates of sponsorship to contributors to the banders' Wind Seine program, picking up slides for the lecture on snowy owls he will give next week, negotiating a group discount at a local motel for the New Jersey Audubon weekend next May, and stopping by the printers to collect the sweatshirts, which now bear David Sibley's drawing of a golden eagle over the CMBO insignia.

At the end of his second fall as director, Kerlinger finds that much of his time goes to such workaday matters. "My father would die if he could see me. He's an academic. I came here as an academic. Now I consider myself a conservationist—a conservationist who knows money makes the world go round. You can't do research without funding, you can't save habitat without political influence, and—" *bam,* he drops another box on the stack—"you can't run a bird observatory without selling sweatshirts."

The computer in the main office is printing out mailing labels. Kerlinger and his staff researcher, Dave Wiedner, have just completed a survey of the economic impact of birding in Cape May, which they are now extending into a study of birders' expenditures across the country. Wiedner distributed questionnaires to birdwatchers at the Point this fall, asking their ages, education, occupations, income, and where and how they spent their money while visiting South Jersey. Kerlinger walks to his desk, reaches between several zigzagging piles of letters, bills, and articles, and pulls out a single sheet of figures. "Eight hundred and thirty-five people responded. They'd come from thirty-one states and six foreign countries. They stayed here an average of five days and four nights, and all told, they used sixty-three different motels, twenty-nine bed and breakfasts, fourteen campgrounds, and ate in a hundred and eighteen different restaurants." Kerlinger estimates ninety thousand birders visited Cape May this year and contributed six million dollars to the local economy. "And the politicians and businessmen of Cape May County hardly realize the birders have been here. As far as they know, the money is a six-million-dollar windfall—unsolicited, unnoticed, and unacknowledged."

Birders need to become vocal about what they are doing, Kerlinger believes, more willing to identify themselves, so people in the local community will recognize the economic value of open spaces and other natural habitat. "It's a three-word message we have to broadcast: *birding means bucks.* Fishermen think nothing of asking a motel manager, 'Where's the fishing been good lately?' And the manager better damn well have the answer, because that's why they're coming to his place. The manager knows he needs those fishermen, and if some local project or development threatens a lake or stream up the road from his motel, he's going to squawk. He has a vested interest in that lake; that lake puts money in his pocket. Birders have to start asking 'Where's the birding been good lately?' and expecting the same kind of attention and respect. Until birdwatchers start asserting themselves like that, so locals will notice the money they're spending in the community, the habitat that should be saved *won't* be saved."

Kerlinger flicks his finger against the page so hard the paper snaps. "This is not a Cape May Point situation. It's nationwide. Birders have not recognized their own economic importance, and they haven't made themselves the political force they could be."

Estimates of the number of birdwatchers in the United States vary widely, depending on how "birdwatcher" is defined. A 1980 census by the U.S. Department of the Interior estimated there were 61 million birdwatchers nationwide, a number nearly equal to the number of hunters (17 million) and fishermen (47 million) combined. A follow-up study, however, determined that fewer than one in fifty of those who enjoyed watching birds could identify more than 100 different species and estimated that "serious birdwatchers" numbered less than a million. Kerlinger believes the second study is misleading. "When you're talking about the economic impact of birdwatching, what matters is not how competent birders are; it's how much money they spend. I run into birders every day in the Meadows, at Higbee's, up on the platform, who can't tell a kestrel from a peregrine at fifty feet. *So what?* They're enjoying themselves, enjoying the natural habitat Cape May still has, and they're spending their time and money in southern New Jersey because southern New Jersey has birds to see. *Birding means bucks.*"

Kerlinger and Wiedner's national survey will be based on responses on a questionnaire they are mailing to three thousand randomly selected participants in National Audubon's annual Christmas Bird Count. The intent of the study is to establish how much birders spend annually on optical equipment, airline tickets, car rentals, natural tours, motels, meals, artwork, bird feeders and birdseed, books, magazines, and memberships. "We know already it's in the billions of dollars, and what we're hoping to do is encourage the birding community to use their economic importance as leverage in their arguments to preserve open areas." Most land-use debates come down to emotion, aesthetics, and long-term thinking on the one side, Kerlinger believes, and economics and short-term thinking on the other. "The conservationists say, 'We have to protect this forest because it's beautiful, and we want to save it for our grandchildren.' But the developer says, 'That forest is wasted space. Let me clear-cut it and put up a hundred condos. You'll increase your tax base when the houses are sold and the new families move in.' The developers win those arguments all the time, even in prime birding areas—because the politicians have to think short-term. But what if conservationists start using economic arguments? 'If he puts up a hundred condos, it's going to *cost* us money. We have to put in new sewer lines, hire more teachers for the school, improve the roads, etc. And right now those woods are *earning* us money. We had ninety thousand birders in Cape May this year. They dropped *six million dollars* in our coffers, and the cost to us in taxes was *zero*. That forest is *not* wasted space. It's very profitable space.' "

The computer stops printing while Kerlinger is brewing himself a cup of tea. He gathers up the labels and carries them, his mug, and a box of manila envelopes over to Pat Sutton's cleared and organized desk, and sits down. He works quickly, peeling and pressing the labels to the envelopes, while he talks. Within five minutes Sutton's desk is buried under the scatter of envelopes and the long roll of empty backing. The mug of tea stands on the corner of the desk, untouched.

Kerlinger has been asked almost every day throughout the season by reporters and other visitors to interpret this year's hawk numbers; he regularly refuses. "A single-site, one-season count doesn't mean squat—any biologist who passed freshman statistics knows that. It doesn't matter what the numbers are, good or bad; you can't account for all the variables. Last year the kestrel flight was two thousand birds better than average; this year it's seven thousand less. Last year I didn't think we could celebrate; and this year I don't think we should panic. There is no such thing as a one-year trend. The peregrine count has been going up pretty steadily for ten or twelve years now. *That* you can call a trend. But one year, good or bad, isn't going to change it. Last fall Jeff counted almost seven hundred peregrines; this fall he's seen three hundred. Does that mean peregrines are crashing? No, the trend is still positive, even though this year shows a drop. You can't keep throwing up the red flag every time you get out in the field and collect some data. Calling wolf will only come back to haunt you—that's another thing conservationists have to learn."

The sharp-shin numbers have him worried, nevertheless. "That's because sharpie is our bread-and-butter bird, and the low numbers this year are beginning to look like they fit into a long-term pattern that we're seeing at other sites too. Sharp-shin numbers have been down since 1984. Maybe it's only a cycle; maybe it's a real decline; we won't know for at least another four or five years, if then. I worry because I hate to think what the Cape May flight would be like if the average count for sharp-shin dropped to five or ten thousand birds a year. The whole feel of this place would change; it wouldn't be the same phenomenon."

The sharp-shin count is especially puzzling in comparison to the apparent upward trend for Cooper's hawks in recent years. Cooper's and sharp-shin are closely related species, and ordinarily considered to fit into similar ecological niches. Both live in deep woods and feed primarily on birds. "But they feed on different sized prey," Kerlinger points out. "And some populations of Cooper's feed heavily on chipmunks. I've seen studies where Cooper's were doing well in one area and declining in another area and the reason seemed to be that one population was feeding primarily on chipmunks, the other only on birds. Maybe we're seeing the same kind of difference between sharpie and Cooper's because of prey size. We know the warbler and vireo populations are crashing—the Breeding Bird Survey data and songbird banding data from almost everywhere show that. If sharpies are depending on those birds more than Cooper's are, that could be a factor."

Another theory suggests the sharpie numbers were artificially high through the 1970s and early 1980s, because they recovered faster than Cooper's hawks from the DDT years and occupied areas Cooper's are now reclaiming. Because no censuses of the coastal accipiter migration were conducted before the CMBO count began in 1976, we may now have an inaccurate sense of what is a normal count for both species and what the ratio of Cooper's to sharpie "should" be.

"I don't accept that theory," says Kerlinger, peeling and pressing. "I'm not convinced the two species compete with each other that much. If we're seeing an effect from sharpies' bouncing back from DDT, it's probably that they boomed for ten or fifteen years while they reoccupied the best of their old territories, and then hit the wall the way any expanding species will. A species can only expand so much before it has to move into peripheral areas—places where the habitat is not quite right, the risks from owls or raccoons are higher, and it's tougher to bring off young. The natural limits to the sharpies' population may be catching up with them now. Maybe Cooper's are still expanding because they were slower to recover from DDT and are still reoccupying their old, prime nesting areas."

A more ominous theory suggests sharp-shins have been stressed in recent years by the loss of woodland habitat all along the Atlantic Coast. Kerlinger believes that first-year sharp-shins follow the coast south because, as unskilled hunters, they need the high density of warblers, vireos, and other songbirds that can be found only along the coast, the primary leading line for songbird migration. These coastal forests and their abundant songbirds have provided a bounty for immature hawks for eons. Kerlinger worries that the effect of loss of these woodlands has been underestimated because our picture of migration is too simplistic and the best-known areas, the migration hot spots, have received the most attention.

"Even the word 'route' is misleading," he says. "When people talk about hawks' following their migration route south, it sounds like an interstate highway—a narrow, well-defined path with exit ramps at Cape May and the other concentration points. The problem with that idea is it suggests that all we have to do is protect the concentration points. I try to use the old term, 'flyway,' because it's fuzzier. Migration is fuzzy. It's a broad-front movement, and migrating hawks are making decisions every moment during flight, in response to weather and their physiological condition. They aren't flying in a simple hop-step-jump from Fire Island to Cape May to Kiptopeke. They are responding each day to that day's conditions, and they're being forced to put down all along the flyway by weather or exhaustion or lack of food. Each individual hawk is in different physiological condition every day and every hour of its flight; and every day and every hour has different winds. The fewer woodlands we preserve, the more stressful we make migration, because we give the birds fewer options. Each piece of woodland we eliminate along the coast is another piece of the safety net cut out."

Kerlinger finds the mug of tea, squeezes out the tea bag and drops it in the wastebasket. He takes a short sip, wrinkles his nose, takes a longer sip, and pauses for a look out the window. "Connected to that problem is the fact that migration research isn't fashionable right now. In some circles of academic gurus it's practically taboo. We know very little about how often hawks feed on migration, how long they can fly between stops, how long they stop over, or how they make decisions about the direction of flight. If we're going to argue to protect stopover habitat along the flyway, we need to address those questions. But the granting agencies aren't funding those kinds of investigations. Migration is too hard to examine in experimental conditions, and right now the sexy, well-funded kinds of projects are controlled experimental studies, where the variables can be manipulated. What's hot right now are things like studies of optimal foraging strategies—someone sets up a series of hummingbird feeders, controls for different amounts of sugar content, bands the birds, and then measures how frequently they come to each feeder. Migration is too damn complicated to study that way. You can't control the variables. I wish we could: 'All right, guys, roll out the cumulus clouds. Let's set the wind at fifteen miles an hour and west-northwest. OK, we're ready. Call New York and tell them to release the hawks.' In migration research you have to let the weather and winds dictate the variables; all you can do is record what you see the birds do."

Kerlinger takes another sip of tea. "What we need more of today is the kind of field research trendy academics think is old-fashioned. Crossley, Wiedner, and Sibley stood on Higbee's dike counting songbirds every morning this year July through October. What did they have? One clipboard, one thermometer, and their binoculars. No white coats, no fancy equipment, and no big budget. But they *quantified* a phenomenon that had never been quantified before. They documented the timing and direction of the flight, the number of birds involved, the species involved, and the seasonality of each species. All we had before this summer were anecdotes. Now we have *data.* We have proven the migration begins in July, that it involves hundreds of thousands of birds, that most of the birds are flying *north,* and that the natural habitat on the Bayshore above the Canal is a crucial buffer zone—an area that hosts thousands of long-distance migrant songbirds almost every day. Now when we argue to protect that habitat, we can provide quantified information to *prove* its importance. That's not old-fashioned, out-of-date research; it's *cutting edge* research."

The phone rings, but the office is closed today, and Kerlinger lets the answering machine respond. He peels and presses labels in silence for more than a minute. "The worst problem in wildlife research is not the trendy academics on the grant panels. The worst problem is that birders aren't bearing their fair share of the money that funds research. Take a look at the budget for the Department of the Interior, and you'll see that most of the revenues that go to wildlife research come from hunters, fishermen, and daily visitors at the national parks and wildlife refuges. Entrance fees generate fifty or sixty million dollars a year. The Dingell-Johnson Sport Fish Restoration Program, which is essentially an excise tax on fishermen, generates something like a hundred and fifty million dollars a year. The Pittmann-Robertson excise tax on hunting paraphernalia and the duck stamp hunters buy generate another two hundred million dollars a year. Hunting and fishing pay the tab so research on ducks and fish gets supported; research on nongame species comes in last.

"Birders will hate me for saying it, but I think we need an excise tax on birding paraphernalia—binoculars, telescopes, bird books, birdseed. And we need a wildlife stamp, which would be the equivalent of the hunter's duck stamp. The question is: who pays for wildlife? If it's hunters and fishermen, then the game species will be protected and the national refuges will be managed for hunting and fishing. If birders want to protect hawks and warblers and shorebirds, they have to start footing their share of the bill."

Brigantine National Wildlife Refuge, forty miles north up the Garden State Parkway, started charging daily entrance fees recently, after years of free admission. Kerlinger thinks it is a good idea. "I heard some birders whining about it, 'It's an outrage. We should boycott the place.' But the ones I respect were all saying, 'Hey, it's about time.' Birders have to see they've had it cheap for too long. I'm sure we'd hear a lot of screaming and fussing if the state started charging entrance fees down here at Higbee's or the hawkwatch platform. But it might happen. If it does and people come in here whining, I'll be ready for them." He turns to face the door as if an infuriated birder has just stomped in. "Read my lips: *No free lunch!*"

*   *   *

By 2:00 P.M. the sky has clouded over, the hawkwatch platform is empty, and neither the Lighthouse parking lot nor the Meadows parking lot holds a single vehicle. The last person watching for hawks today seems to be Al Nicholson, who is packing up his painting equipment at the Beanery.

Nicholson's season is ending also. "I usually quit painting by the end of November. The leaves are down, and you don't get the good skies any longer. Maybe once or twice on a warm December day you get something interesting, but usually the sky is too sodden."

He works with oils, standing at a tall easel, blocking out the trees and the field or woods he faces, then painting the sky, working primarily with a palette knife, as the clouds brighten and shift. Generally, he has three or four paintings in progress at once, and chooses which he works on as the conditions change. "The sky is the

Pond Creek

organ of sentiment. You have to keep things fluid, and you have to be ready for anything. People don't know how hard painting is. The day before yesterday I had a really good day. I surprised myself. That's twice this fall that's happened. Birdwatchers think warblers are exciting. When you're painting well, it's *exhilarating*. There's nothing like it."

Two weeks ago, on November 11, David Sibley found a Townsend's warbler down the railroad tracks two hundred yards from Nicholson's easel. It's a western species, rarely seen east of the Mississippi, and birders trooped by Nicholson for a day and a half while the bird remained feeding in a small grove of trees. Nicholson never left his position to go see it, and most of the birders veered around him. He is a man few know. Only Sibley, a fellow painter, and Clay Sutton, Nicholson's old friend, stopped to talk.

Today, however, with the clouds overhead, Nicholson has

decided he should walk the woods. He points to a red-tailed hawk
in a tall dead oak. "He was here yesterday and all day today, looking
this way." He kneels to put away his paints and slide his mixing
boards into a paint box that is so warped the clasps will not close.
He has had it since his art school days at the Pennsylvania Academy
of Fine Arts and the Tyler School at Temple University, in the 1950s.
"I keep it to give my critics a laugh. 'My God, look at his box.'"
The original handle is gone, replaced by a couple of inches of plastic
rope, and Nicholson must hold the sides together with one hand as
he opens the trunk of his car. The car, a bronze Toyota, is filled with
rags, jars, and tubes of paint. Nicholson puts his painting upright
between two other canvases behind the passenger seat, then folds his
easel and slides it into the recesses of the trunk.

He retrieves his binoculars from the trunk and zips up his
coat, which is elbow-less, spattered with paints, and reeks of turpen-
tine and linseed oil.

Nicholson strides westward along the railroad tracks, then
turns north to Pond Creek. Pods of corn in the field to the south rattle
in the wind. The breathy squeaks of white-throated sparrows come
from every bush, but he does not stop to look for them. Of Cape
May's other birders only Katy Duffy walks at a pace that matches
Nicholson's. Most others walk exceedingly slowly, inspecting each
tree and bush for birds. Nicholson's walk is a long-striding charge,
crashing through briers. His intention seems to be more like a photographer's than a birder's—to collect as many images at as many different angles as quickly as possible.

He points to a turkey vulture beating north over the trees
along Bayshore Road. "You can't see that from the hawkwatch. I
don't mean the bird. I mean the view. The hawkwatch is blank. Here
you see the tones. That's what's interesting. The tones, the light, and
the conditions."

Season's End

❖

Nicholson grew up in Moorestown, New Jersey, about ten miles east of Philadelphia, and eighty miles north of Cape May. His interests in art and in raptors have been inseparable all his life. He cannot remember when he was not fascinated with drawing and with birds. "I used to copy Audubon's prints. His snowy owl print. His Richardson's owl." One Christmas his father bought him a mounted red-tailed hawk from a taxidermist in Medford. Nicholson drew it over and over again. "I was fascinated. It was a wonderful thing."

He explored the local and not-so-local woods by bicycle. "I was born right at the tail end of what was. I used to bicycle to Parker's Creek outside Moorestown, a wonderful cattail meadows with peregrines and rough-legged hawks and short-eared owls, and I'd go over to Riverton on the Delaware River. There was still a vast long-eared owl roost there then. One day I flushed forty long-ears from a tree. People who live in those places today don't know that background. You have to be really old."

Through the Quaker Friends School he attended in Moorestown he was introduced to a circle of men who knew birds and Witmer Stone and who traveled to Cape May regularly: William Evans, Albert Linton, Julian Potter, Richard Miller. "I was wild for any contact with these people. They could show me things. My mother would have William Evans to tea just so I could talk to him and learn from him."

The Quakers went by train to Cape May each winter in the 1930s for the Christmas Bird Count. "The bird census was the most inspiring thing—to come down here when it was still a wilderness. We'd go out to North Wildwood and see snowy owls on the beach, walk through the bayberries and find short-ears in the grasses. One year we found a Barrow's goldeneye on the ocean. It was a wonderland.

"World War Two set everything back because it sped life

up so. Housing developments, heavy equipment, logging, all of them were damaging. Until the logging started during the war, all the wetlands in Cape May County and Cumberland County were intact. Bear Swamp was intact. Bald eagles nested in the cedars in Goshen. And Cox Hall Creek then was the most beautiful place I've ever seen in the whole world. Soaring eagles, peregrines looping through the sky, and ospreys—*endless* ospreys."

Nicholson served in Air Force intelligence in Korea and in London, and came home to what he feels was the end of the old Cape May. The completion of the Garden State Parkway in 1955 was a crucial blow, Nicholson believes, because it ended Cape May's isolation from the rest of the state. "The Parkway changed things forever. It modernized Cape May County. Nowadays things are so bad the Parkway has become a conservation force, because it has tied up a lot of areas where salt marsh and wetlands meet the forest, and protected them from further development. But in the 1950s the Parkway decimated those places. The center of Rio Grande was a huge forest. Now it's a shopping mall."

Nicholson's favorite retreat during the 1950s was the South Cape May Meadows, then an oceanside cattle pasture. "I went there every day, trying to teach myself to paint. It was a vast place then, stretching to the sea, and teeming with life, *seething* with life—snipe, egrets, terns, waterfowl, glossy ibis. Every fall there were big flights of upland sandpiper, and their call notes came down from the sky like rain. You might hear them now, I suppose, from the hawkwatch. But it's different when you're the only one listening—when it's a private imaginative world. I didn't have much money then, but it didn't matter. I had no material ambitions, I just wanted to study things, and the Point was a magical place.

"It's too crowded down here now. The value, the tremendous spiritual impact of the Meadows and the other natural areas, has vanished. The present generation of birdwatchers will never know them as they were.

"It's the same everywhere in this country. There's no longer a frame of reference whereby there can be a determination of what's right and what's wrong—what we need to do and what we don't. Orlando, Florida, used to be an eagle roosting area; you could see eagles by the dozens there—soaring and nesting, living their lives. It was a magnificent wetlands. Now it's Mickey Mouse Land. Bush gets himself photographed there shaking hands with Mickey Mouse, and people like that. They *vote* for him. They think it's great. The television world, Mickey Mouse Land, is all people know today. Real contact with nature, spiritual contact, is gone."

Six steps after making this remark, Nicholson trips on a rusted line of barbed wire, falls on his face, and rolls over laughing. "I skin my shins on that *every* time I come in here."

For all his doomsday talk, Nicholson is not a dispirited man. Walking with him through the Cape May landscape is like walking with an archaeologist through Athens or Venice. The glory that is gone accentuates the beauty that remains. And like the archaeologist, Nicholson treasures everything that is old. The older the better. He points to an oak tree as thick as a stone column. "That's old. That's really old." Twenty yards farther he finds another, still thicker oak. "Now *this* is really old." He says, "Isn't it wonderful how time goes on back in here?"

Even abandoned machinery earns Nicholson's eye. "Look at that," he says, pointing at a crane, wheels missing, which is rusting away alongside the railroad trails. "That's old. That's really old. A walk back here is a journey in space and time."

He kneels to examine the barely discernible remains of Old Clamshell Road, which is more than a hundred years old and seems to have run from the Shunpike to the Point. Today, it is only a rise in the woods, covered thickly by oaks and maples. Not far away, Nicholson points to a duck decoy floating low in Pond Creek. The paint is still bright, but the wood has become logged with water, and the bird is up to its neck. "That's been back here for years," he says. "It still fools me every time."

He pushes through a line of cedars to examine the Creek where the phragmites are encroaching on the cattails, another sad invasion of a man-assisted alien over a native species, like the starlings' replacement of the bluebird and the house sparrows' replacement of the cliff swallow. Like the bluebird and cliff swallow, cattails are hard to find in Cape May today. "You can't do anything about it," says Nicholson. "Just leave them alone. That's all you can do."

A hundred yards later he pushes through another line of cedars to view another section of the Creek, where the cattails are ten acres strong. "Now, isn't this great? *This* is how a pond is supposed to look. Right here is where I saw my first ferruginous hawk. Over those trees, soaring. It was a November day with puffy, magnificent clouds, climbing into the heavens. I don't remember the year."

He checks the sky every thirty seconds, not only for birds, but for breaks in the clouds. In Nicholson's view, the two are linked: changes in the light foreshadow the coming of the hawks. The sky is unpromising now, however; the sun is a diffuse glow under a thick gray, like a dying flashlight under blankets. "You never know what's going to happen," he says again. "Sometimes things just open up."

At the southernmost bend of Pond Creek, he pushes through ten yards of phragmites to point to a view of the forest and the stack of the old magnesite plant. Despite the decades of pollution it created, the plant, in its derelict state, seems somehow to have earned its way into Nicholson's heart. "Oh, now there's a view," he says. "You can't see *that* from the hawkwatch."

He turns and heads back, at the same hard pace, and points

to a Cooper's hawk perched almost out of sight near the trunk of a maple. Amazingly—in violation of all norms of accipiter behavior—the bird doesn't move as Nicholson tromps its way, even when he is close enough to view the hawk's orange eye through his binoculars. "Back in here, he feels safe," Nicholson explains. "In here deep is where he belongs."

Nicholson stops for another view of a cattail expanse on Pond Creek. He pushes an overhanging branch out of his face and scans the tree line for hawks. "Two golden eagles hung around the Pond for a long time one winter in the late fifties, feeding on the ducks. One afternoon I was standing right here, looking up, when one of them dropped on the ducks. He was a streak, a bolt, and the wind through his feathers sounded like trees crashing in the forest. This was back in the days when nobody believed golden eagles occurred in Cape May. Even my mother didn't believe me."

He turns to go, spots something off in the woods, studies it with his binoculars, bends to crash through the trees, stops to relocate it with binoculars, crashes another few steps, then finally kneels above his discovery. "Isn't that wonderful?" he asks, pointing. It is a fungus the color of burgundy wine and the size of a tennis ball, which clings to a fallen log.

When Nicholson emerges from the woods again, two hours

after he closed up his easel, the red-tailed hawk stands in the same dead oak. "That was a good walk, and our friend is still here," he says, lifting his binoculars to the bird.

Suddenly, Nicholson steps backward as if he has been shoved and lowers his binoculars. "Oh, *look* at that," he says, pointing. A mile above the hawk the clouds are breaking; shafts of sun stream downward. "Just *look* at that!" Nicholson shouts, taking another step backward and throwing his arms as wide as a preacher's. "Isn't that magnificent? That light-torn sky! That fluffy agitation!"

Afterword

❖

NOW THAT IT'S DONE, THIS BOOK SEEMS TO HAVE BEGUN ON THE morning of October 15, 1978, when I visited the Cape May Point Hawkwatch and met Pete Dunne for the first time. I was in a glum and cynical mood. A job change had forced me to leave Florida, where I had taken up birdwatching, and move back to New Jersey, where I had grown up and which I thought of as the most ordinary state in the nation. *American Birds* had just published its account of Dunne's 1977 census and granted Cape May its new title as "the Raptor Capital of North America," and I had decided this was some kind of trick Dunne had pulled on the editors. No one had exposed the fraud, I guessed out of my ignorance and New Jersey inferiority complex, because no serious birder could be bothered driving all the way to the bottom of such a humdrum state to spend a day at Dunne's side. The most incredible of Dunne's numbers was his count for peregrine falcons, a bird I thought of then, as I still do, as the epitome of animal otherworldliness. In four full years of diligent birding in exotic, still-wild Florida I had seen exactly two peregrine falcons; in two fall seasons in mundane, hopelessly industrialized New Jersey Dunne claimed to have seen 166. It just couldn't be true.

I arrived that first day to find him perched on a rung of his lifeguard chair, surrounded by a half-dozen birders, and even before the first hawk came by I knew Dunne wasn't making anything up. The intensity and energy around that chair was like no birding I'd ever witnessed. I felt like an intramural softball player who had sneaked into the dugout of the New York Mets during a pennant-deciding game. But Dunne drew me in. "Good morning," he said. "Where are you coming from today?" "Florida," I told him, over-simplifying. "You've got some birds on me, then," he said. "I've never even seen an anhinga." This was the equivalent of Tom Seaver admitting he had never learned to pitch underhanded, but it allowed me to step forward into the circle of watchers. A peregrine soon flew by, and half a minute later along came another. "Back to back," said someone. "As usual," said someone else. Half an hour after that, Dunne pointed out the first rough-legged hawk of the season cresting the horizon, and as the bird soared over the group I couldn't resist a whoop and a little dance of excitement. Dunne smiled, eyebrow arched. "I guess that was one bird I had on you, no?"

I returned to the Point again the next weekend, and I have been birding there as often as possible ever since. In the fall of 1988, ten years after my first visit, I managed to spend enough days at the Point, thirty-four of the one hundred days of the hawkwatch season, to feel that the details and chronology of that single season might make a frame for the larger story.

In the two counts conducted since the one described in this book, peregrine falcon numbers have continued to be higher at Cape May Point than anywhere else in the world. In 1989 the count was 703 peregrines; in 1990 it was 818, a new all-time record. Cooper's hawks and ospreys have also been seen in numbers well above their average over the first ten years of the count and higher than anywhere else in the world. Sharp-shins and kestrels, on the other hand, have been down. In 1989 only 10,625 sharp-shins and 7,185 kestrels were recorded, and the total of all hawks was 29,737, the lowest since the CMBO count began in 1976. Last season, 1990, the total was a couple of hundred hawks lower still—29,406—and the low counts in sharp-shins and kestrels, 12,495 and 7,345, were the primary reason. Most other species were seen in average or higher than average numbers.

"Let's see what happens next fall," everyone says now—as everyone says every winter.

Almost all the birders named in this book have remained at the Point or returned there for the fall migration. Jeff Bouton has spent both fall seasons in the banding stations and has worked under Chris Schultz in Colorado in the spring and summer, banding nestling peregrines. This coming spring he will direct the hawk- and owl-banding operation at the Braddock Bay Raptor Research Center on Lake Ontario. Chris Schultz continues to direct CMBO's hawk-banding project each fall, and on October 16, 1989, he had the great thrill of trapping a banded peregrine at Far North Station, checking the number, and realizing it was a bird he had banded himself in Greenland, seventy-eight days earlier, three thousand miles north. (He remembered it well: a female, about two weeks old when he last had seen her, the youngest and smallest of four nestlings in an aerie above a small lake called Ring So in west central Greenland.) David Wiedner now works in the Philadelphia Academy of Sciences, but visits the Point regularly. Richard Crossley has been in Japan for the past two years, but he is expected back at the Point for this spring's migration, and rumor has it that he may move to Cape May permanently. Vince Elia and Bob Barber continue to walk the woods of Higbee's Beach and the Point's other birding spots every weekend year-round. Elia's luck with orange-crowned warblers changed soon after the day described in Chapter Eight, and on April 20, 1990, he discovered a cave swallow over Bunker Pond, the first ever identified in New Jersey and one of the very few ever recorded anywhere in North America north of Florida. Katy Duffy continues to direct the owl-banding operation, and Pat Matheny, Joey Mason, Kim Stahler, and most of the other banders mentioned have also returned to work in the blinds. Paul Kerlinger remains as director of CMBO, speaks and publishes regularly on the economics of birding and on the songbird migration at the Point as well as other topics, and has successfully solicited funds for a number of research projects and for a new headquarters. (Membership and other information is available from The Cape May Bird Observatory, P.O. Box 3, Cape May Point, NJ 08212.) Al Nicholson still paints at the Beanery. Clay Sutton continues to chart the hawk migration diligently. In 1990 he conducted an "alternate hawkwatch" at East Point up the Delaware Bayshore, coordinating his count with the platform count and watching for hawks the banders had marked with plastic tags. Though the flight at East Point seems significant—apparently totaling about a third the number of hawks seen at the Lighthouse—Sutton saw not a single tagged bird, so the direction of the flight of the hawks after they leave the Point remains a mystery. Pat Sutton continues her hard work on a number of conservation fronts and earned praise for her efforts from all sides when the new Cape May National Wildlife Refuge was dedicated in May 1989. If land acquisition proceeds as expected, it will eventually comprise 14,000 acres of precious wildlife habitat on the Delaware Bayshore.

A problem that seemed secondary two years ago has become a primary worry for the birders and naturalists of Cape May. The beach erosion at the Point seems to have accelerated in recent months. The bunker is now inaccessible and completely surrounded by water, and the South Cape May Meadows and Bunker Pond appear to be in imminent danger of destruction. "Just one bad storm," people say, shaking their heads, "and the dunes will be gone." The state of New Jersey has trucked in sand in past years to protect the Meadows and the Park, but money is tight now and there seems to be a sense among state authorities that any further action will only postpone the inevitable. If you want to see the Cape May flight from the platform as described here, come soon.

I am grateful to Stockton State College for the research grant that enabled me to begin work on this book. I am grateful also to each of the people who appear in the text for their willingness to share their words and their lives. Any errors, misstatements, or distortions are my responsibility. Jeff Bouton especially deserves mention for bearing with my incessant questions (and many misidentifications) through the three-and-a-half months of the season—and for his help during the writing of this book. Paul Kerlinger, too, deserves special mention: he supported this project even before I was able to explain it to him; and he has been helping me arrange interviews, track down information, interpret terminology, and find places to sleep ever since. I cannot express my gratitude for the hospitality and friendship of Pat and Clay Sutton: both have taught me more about birds, the natural world, and conservation ethics than they can know.

Others whose help, encouragement, and expertise kept me working on this project include Don Almquist, Hoong Chang, Bill Clark, Carson Connor, Robert Connor, Jeff Dodge, Castle Freeman, Peter Knapp, John Kricher, Allen Lacy, Janis Nazarenko, Roger Tory Peterson, Tom Rawls, David Sibley, Mitch Smith, Nat Sobel, Heather Stohr, Mary Anne Trail, Dick Turner, Dave Ward, and Louise Zemaitis. I am especially grateful to John Barstow, my editor, whose patience and wisdom have made this a much better book than it could ever have been without him.

AFTERWORD

❖

Thanks to Blake, Colette, and Teal for bearing with my absences and for cheering me on.

Finally, the loudest and most heartfelt thank you goes to Jesse Easton Connor—for her kindness and her support over all these years.

JACK CONNOR
Port Republic, N.J.
February 1, 1991

Appendix

❖

A SHORT GUIDE
TO CAPE MAY'S
HAWKS

HAWKWATCHERS' TERMINOLOGY IS LESS COMPLICATED THAN IT MAY seem at first. The most loosely used word is *raptor,* a term that applies to both hawks and owls. Both groups of birds have hooked bills, strong legs, taloned feet, and feed primarily on live prey. Hawks are sometimes called *diurnal raptors* since no hawk hunts at night. Despite the birds' common name, however, taxonomists believe hawks and owls are not closely related, and classify all hawks as members of the order *Falconiformes.* The order comprises about 285 species worldwide, from the tiny, shrike-sized falconets of India and Africa to the harpy eagle of South America, an enormous bird which feeds primarily on monkeys and sloths. Thirty-two species of *Falconiformes* breed in North America: the osprey (a unique species not closely related to other *Falconiformes*), the northern harrier (the single North American representative of a worldwide genus), two vultures, five kites, three accipiters, twelve buteos, two eagles, and six falcons. Nineteen of these thirty-two species are regular migrants in Cape May. Five other *Falconiformes* have occurred in Cape May accidentally.

❖

Osprey, *Pandion haliaetus:* The "fish hawk," a widely distributed species nesting in Southeast Asia, Japan, Australia, and parts of Africa and the Middle East, as well as throughout the United States. With its black-and-white plumage and kinked-wing glide, the osprey is more likely to be mistaken for a gull than a hawk. It feeds exclusively on fish, which it catches in diving descents from as high as 150 feet above the surface. It is a midseason migrant at Cape May, peaking in early October, and totaling about two thousand individuals a year.

Northern Harrier, *Circus cyaneus:* Also known as the "marsh hawk"; a long-winged, long-tailed, medium-sized raptor, best identified by its white rump patch and by its characteristic hunting style—a slow, coursing flight low over marshes and other open areas, where it feeds primarily on mammals. On migration, it flies at all heights, gliding frequently, and with its swept-back wings and squared tail, it can be mistaken for a falcon. The harrier has the most extensive migration period of all Cape May's raptors: the first southbound in-

dividuals, usually immatures, appear in early August, and the last, usually adult males, can still be seen flying south in mid-December. The average seasonal count at the Point is about eighteen hundred individuals.

Vultures are large, dark-plumaged, bare-headed scavengers that feed on carrion, which they search out while soaring. Taxonomists now believe that neither North American vulture is a true member of the *Falconiformes* and that both are more closely related to the storks. Traditionally they have been considered raptors, however, and are still included in most hawkwatch censuses.

Turkey Vulture, *Cathartes aura:* A year-round resident of Cape May and difficult to census because nonmigrants are visible from the platform almost every day, gliding on wings held in V-shaped dihedrals. Migrants move through the Point from August to December and total about six hundred individuals each fall.

Black Vulture, *Coragyps atratus:* Primarily a southern species, only recently nesting in New Jersey. Generally one to three individuals are seen each year at the Point, most often in November.

Kites are long-winged, small-bodied hawks, named for their buoyant, windblown flight. The two species that occur in Cape May feed primarily on insects, especially dragonflies, which they pluck from the air with their talons and tear apart on the wing. Neither is a regular autumn migrant.

Mississippi Kite, *Ictinia mississippiensis:* A small, graceful raptor that nests from the Carolinas south and west to Arizona and winters in Central America. Five to ten individuals (apparently overshooting northbound migrants) stray to Cape May annually in late spring; the species is much rarer in fall, with only three or four accepted autumn records.

Swallow-Tailed Kite, *Elanoides forficatus:* A beautiful, fork-tailed hawk that nests in scattered swamps and river bottoms in Florida and along the Gulf Coast and winters in South America. One to three individuals stray each spring to Cape May; a sighting in 1946 is still the county's single fall record, however.

Accipiters are secretive, forest-dwelling raptors with long tails and rectangular, relatively short wings that prey primarily on birds they capture in dashing, zigzagging chases through branches and bushes. On migration and in open situations, they fly with a distinctive herky-jerky rhythm: flap, flap, glide, flap, flap. Like owls, they usually perch out of sight, hidden behind branches and leaves. All three North American accipiters occur at Cape May; from smallest to largest they are:

Sharp-Shinned Hawk, *Accipiter striatus:* Cape May's most numerous hawk and one of the most common raptors in North

America, although its secretive habits and lack of distinctive field marks make it an unfamiliar species to casual observers. Birders generally consider distinguishing sharp-shins from Cooper's hawks the most difficult identification puzzle among North American hawks. Both species have brown backs and streaky, brown-and-white ("dirty") breasts as immatures and have blue-gray backs and rufous breasts as adults. Their silhouettes, flight styles, and habits are also very similar. Sharpies are slightly smaller, however, and show a square-cut tail, usually with less white at the tip. They are midseason migrants at Cape May, peaking in early October. The Point's average count for sharpies has decreased dramatically in recent years. From 1976 to 1985, the average was more than 40,000 a season; in recent years it has been less than 25,000.

Cooper's Hawk, *Accipiter cooperii:* The sharp-shin's sister species (see above). Cooper's are slightly larger than sharp-shins and have more rounded tails, which often show a wide band of white at the tip. Like sharpies, they feed most often on birds, but certain populations apparently prey regularly on chipmunks and small mammals. Cooper's hawks have been increasing at

the Point since the late 1970s and have averaged nearly 2,000 individuals a season in recent years. The peak period is early to mid-October.

Goshawk, *Accipiter gentilis:* The most northerly nester of the accipiters, and an erratic migrant at Cape May, coming south in good numbers only about once every ten or twelve years, apparently when prey populations decline in Canada and New England. Goshawks are large and powerful hawks, nearly the size of red-tails, and feed on ducks, grouse, and rabbits as well as smaller birds. In 1985, the most recent invasion year, eighty-seven goshawks were counted from the platform; generally the count is between fifteen and thirty a season, mostly immatures in late October and November.

Buteos comprise the most diverse and successful genus of hawks in North America. All are broad-winged, wide-tailed, soaring birds, whose in-flight silhouettes resemble those of crows or ravens. Most buteos hunt from the wing, feeding most often on rats, mice, voles, and other small mammals. In increasing size Cape May's five regular buteos are:

Broad-Winged Hawk, *Buteo platypterus:* The smallest eastern buteo, and the most erratic migrant of the five most common hawks at Cape May. It can be distinguished by its size, its broadly banded tail, and its flocking habits. Though a long-distance traveler, which nests from Florida to Canada and winters in Central and South America, the broad-wing is a very reluctant water-crosser. Thus, it is more common on the inland ridges than at Cape May or elsewhere along the coast. The broad-wing is an early-season migrant at the Point, peaking in September, but its numbers vary widely each autumn, apparently in response to winds and weather conditions, from low counts of less than a thousand for the season to high counts of more than ten thousand.

Red-Shouldered Hawk, *Buteo lineatus:* A midsized buteo which is occasionally mistaken for a goshawk because of its flap-flap-glide flight style and its rectangular, accipiterlike wings. The best field marks are the translucent crescents (the "windows") in the outer wings, the red breast and shoulders of the adults, and the reddish wing linings on both adults and immatures. Red-shoulders are shorter-distance migrants than broad-wings, wintering primarily in the southeastern United States, and come later in the season to Cape May, peaking in late October or early November. The species is apparently on the decline in many parts of its range, but numbers have been fairly steady at Cape May, averaging about 500 individuals each autumn.

Swainson's Hawk, *Buteo swainsoni:* A western species that was once considered extremely rare east of the Mississippi but is now regularly reported in the East, especially at Cape May where three to five individuals are seen annually, usually in September or early October. Swainson's are slightly smaller than red-tailed hawks, have longer, narrower wings and, in the more common light-phase plumage, show a dark bib under the throat. Most of the population winters in Argentina, but a few immatures straggle to southern Florida each year.

Rough-Legged Hawk, *Buteo lagopus:* The most northerly of buteos, nesting in boreal forests and on the tundra above the tree line; a late and erratic migrant at Cape May, coming south in numbers only once every three or four seasons, apparently when lemming numbers and other prey populations are at cyclic lows. Rough-legs are nearly the size of red-tails but have trimmer profiles, slightly longer wings, and entirely dark bellies. The average autumn count at Cape May is about five individuals, most often in November and December.

Red-Tailed Hawk, *Buteo jamaicensis:* The largest, most widespread, and generally most visible of buteos in the eastern United States. The red-tail perches regularly on telephone poles and other manmade structures and can be found even in urban

areas. In the East adults have orange-red tails, and immatures have brown, finely striped tails; both age classes can also be identified by the narrow dark band crossing the belly halfway between feet and throat. The species is a late migrant at Cape May, with numbers peaking in November and usually totaling between fifteen hundred and two thousand for the season.

Eagles are huge raptors, with large heads, heavy and prominent bills, wide and rounded tails, and extremely large wings (six to seven feet from tip to tip). In flight, they look like oversized buteos, soaring with their wings held flat on the horizontal plane and their primary feathers splayed.

Golden Eagle, *Aquila chrysaetos:* A regal and breathtaking raptor, and the bird many hawkwatchers wish had been named our national emblem. The golden eagle is a superb flyer—swift and powerful—and a masterful predator, preying regularly on mammals as large as prairie dogs and jackrabbits and also on birds, including other raptors such as the goshawk, red-

tailed hawk, and short-eared owl. It is generally darker in immature plumages than the immature bald eagle, and is all dark as an adult, with a golden head and nape. It nests very sparsely in North America in high and open country primarily in the West and Canada. About ten individuals are seen each autumn at Cape May, most often from mid-October to late November.

Bald Eagle, *Haliaeetus leucocephalus:* Our national emblem is primarily a scavenger in the East, feeding most often on dead or dying fish and carrion. It is slightly bigger than the golden

eagle and has a more prominent head and bill. The bald is generally an earlier migrant at Cape May than the golden, first appearing in July and peaking in late September. Numbers have been increasing in recent years, as the species recovers from DDT. For the first ten years of Cape May's count the average was less than twenty a season; in recent years it has been more than forty. Most bald eagles seen at the Point are immature birds.

Falcons are small to medium-sized hawks that are shaped like giant swallows, with swept-back, pointed wings, prominent chests, and a swift, graceful flight. They are open-country birds; the larger species feed on birds they overtake on the wing and sometimes kill on impact. Three species are regular migrants through Cape May; in increasing size and decreasing abundance they are:

American Kestrel, *Falco sparverius:* Like the red-tailed hawk, a familiar raptor throughout the United States because it is so numerous and perches so frequently in the open on manmade structures—especially fences and telephone lines. Both sexes can be identified by their small size, red backs, and the two

"teardrop" or "mustache" marks on each side of the face; adult males also have blue wings. Unlike the larger falcons, the kestrel feeds on insects and small vertebrates more often than birds and hunts from a perch or hovers in flight before dropping on its prey. Ten to fifteen thousand kestrels are counted from the Point's platform most falls; it is the second most common hawk at Cape May.

Merlin, *Falco columbarius:* A small, dark-breasted falcon with a pumping, swift, and nonstop flight style; a.k.a. "the bullet hawk"; slightly larger, much less common, and a more localized and northerly nester than the kestrel. It is a feisty and combative species, feeds regularly on birds as large as flickers, grackles, and meadowlarks (which equal it in size); has been observed killing chickens, pigeons, and ptarmigans (which exceed it in size); and can be identified at long range simply by its willingness to harass all other raptors, even including golden eagles. The average count per season at Cape May is about fifteen hundred individuals, the majority in October.

Peregrine Falcon, *Falco peregrinus:* The most widespread raptor in the world, nesting on all continents and many islands in both hemispheres, but generally rare and localized throughout its range. Adults of both sexes have white breasts and a "hooded" look with dark caps extending down past the eye into a single "mustache" mark on each side of the face. Immatures have streaked breasts and lighter crowns. All peregrines have an elegant and fast flight, with lashing wing strokes and swallowlike glides. Their favorite prey are doves, ducks, and shorebirds, which they catch on the wing and generally kill on impact. North America has three native populations—a nonmigratory population on the West Coast,

a midcontinental population that nests in high-mountain areas from Colorado north to Alaska, and an arctic race that nests above the tree line across Canada into Greenland. Recently birds that have been bred in captivity have been introduced along the East Coast and in some urban areas. The peregrines seen at Cape May are believed to be primarily arctic birds heading south to winter in the Caribbean and South America. Cape May has averaged more than 400 peregrines a season in recent years, more than can be seen anywhere else in the world.

Accidental Species

European Sparrowhawk, *Accipiter nisus:* One photographed from the platform in October 1978; three or four other reports from Cape May, all disputed. The closest breeding grounds are in Ireland.

Ferruginous Hawk, *Buteo regalis:* A buteo ordinarily restricted to the plains and other open country from the Dakotas west to

California; about half a dozen fall reports exist for Cape May County, some disputed.

European Kestrel, *Falco tinnunculus:* One bird banded and photographed at the Point in 1972; one or possibly two individuals identified on the wing in 1979. Fewer than a dozen records of this Old World species exist from elsewhere in North America.

Eurasian Hobby, *Falco subbuteo:* Cape May's 1986 record of this bird is one of fewer than half a dozen reports of the species in North America. It breeds from Great Britain west to China, and winters in Africa, India, and Southeast Asia.

Gyrfalcon, *Falco rusticolus:* An arctic raptor, the largest and rarest of North America's nesting falcons; there are several winter records for Cape May County, but one gyr reported on October 14, 1983, is the only fall record.